D0296343

Some thermodynamic aspects
of inorganic chemistry

Cambridge Chemistry Texts

GENERAL EDITORS

E. A. V. Ebsworth, Ph.D.
Professor of Inorganic Chemistry,
University of Edinburgh

P. J. Padley, Ph.D.
Lecturer in Physical Chemistry,
University College of Swansea

K. Schofield, D.Sc.
Reader in Organic Chemistry,
University of Exeter

TO JOC

Some thermodynamic aspects of inorganic chemistry

D. A. JOHNSON

Fellow of Trinity Hall, Cambridge

CAMBRIDGE

at the University Press 1968

Published by the Syndics of the Cambridge University Press
Bentley House, 200 Euston Road, London, N.W.1
American Branch: 32 East 57th Street, New York, N.Y.10022

Library of Congress Catalogue Card Number: 68–29118
Standard Book Number: 521 07108 9

Printed in Great Britain
at the University Printing House, Cambridge
(Brooke Crutchley, University Printer)

Preface

This book is designed as a supplement to the more comprehensive works that are used by students taking a university course in inorganic chemistry. I have assumed that readers are acquainted with the elementary thermodynamics of chemical equilibria.

Thermodynamics may be applied to chemistry in an interpretive role when the energies of chemical reactions can be explained with the help of theoretical models. Thus the value of the thermodynamic approach is limited by the power of current theories. For this reason, the emphasis in the book is rather different from that of a modern descriptive text. Chapters 2, 3, 5 and 6 are concerned mainly with simplified ionic models, and with the compounds or ions to which they can be applied. Only chapter 7 contains any detailed discussion of covalent substances.

There are a number of people whose help I should like to acknowledge. By their disinterested award of a research fellowship, the Master and Fellows of Trinity Hall, Cambridge furnished me with the independence that I needed to write the book. I should like to thank them for the individual kindnesses that they showed me and, with the exception of one or two convivial historians, for concealing any dismay at my dilatory progress.

Professor E. A. V. Ebsworth provided me with a great deal of patient encouragement while Dr D. A. Haydon, Dr D. J. Miller, Dr A. G. Robiette and Dr R. Walsh all suggested valuable alterations to certain sections of the text. Both Dr A. G. Sharpe, who by some stimulating undergraduate and graduate supervision first introduced me to the subject of this book, and Dr J. J. Turner, read the greater part of the manuscript at different stages and made a number of very helpful criticisms.

During the preparation of successive drafts, much my most generous helper was Dr P. G. Nelson, Lecturer in Inorganic Chemistry at the University of Hull. The treatment of d^n ionization potentials by separation of the coulombic and exchange terms is taken from his Ph.D. thesis, and it was he who suggested to me the idea of differentiating functions of the simple Kapustinskii equation. He made many other very valuable contributions in the absorbing hours of discussion that were spent while I enjoyed his hospitality at Hull. As these are too numerous to mention, perhaps the best thing I can do is to express my admiration of his mastery of chemical exposition and of his profound understanding of the subject.

None of the friends and colleagues that I have mentioned is to blame for the errors or imperfections in the book. For these I take full responsibility.

D.A.J.

15th March, 1968

Contents

Contents

1. Introduction

1.1. Stability. Among other things, inorganic chemistry tries to account for the differing stabilities of different chemical systems. A system is said to be stable if it does not appear to change with time. Physical chemistry draws a sharp distinction between real and apparent stability. In the first case the system is in a state of equilibrium and it is stable in the strict sense of the word: that is, none of the conceivable changes in the system can occur spontaneously. In the second case the system is not in a state of equilibrium and it is only apparently stable: that is, at least one of the conceivable changes can occur spontaneously but it does so at an immeasurably slow rate. The first type of stability is studied by that branch of physical chemistry known as thermodynamics; the second by that called reaction kinetics.

1.2. Thermodynamic stability.[1] For any chemical reaction,

$$a\mathrm{A} + b\mathrm{B} + c\mathrm{C} + \ldots \rightarrow k\mathrm{K} + l\mathrm{L} + m\mathrm{M} + \ldots, \qquad [1.1]$$

the position of equilibrium is conveniently expressed in terms of the equilibrium constant K, a function of the activities, a, of the reactants and products:

$$K = \frac{a_{\mathrm{K}}^{\mathrm{k}} \cdot a_{\mathrm{L}}^{\mathrm{l}} \cdot a_{\mathrm{M}}^{\mathrm{m}} \cdot \ldots}{a_{\mathrm{A}}^{\mathrm{a}} \cdot a_{\mathrm{B}}^{\mathrm{b}} \cdot a_{\mathrm{C}}^{\mathrm{c}} \cdot \ldots}. \qquad (1.1)$$

For pure liquids and solids at normal pressures, it is a very good approximation to put the activity equal to one. For solution components, whether the continuous phase be solid, liquid or gas, the activity can usually be replaced, without severe error, by the concentration if the solutions are fairly dilute.

For the application of thermodynamics to descriptive chemistry,

[1] Thermodynamic properties in this book are quoted at 25° unless otherwise stated. Values of fundamental constants are from Rossini (1964).

an equation relating equilibrium constants to chemical energy is essential. This is provided by the van't Hoff isotherm

$$-\Delta G^0 = RT \ln K. \tag{1.2}^1$$

ΔG^0 is the standard free energy change of the reaction to which K refers. In tables of thermodynamic data, it is the standard free energy of formation of a compound, ΔG_f^0, that is usually recorded. For any balanced reaction, at constant temperature,

$$\Delta G^0 = \Sigma \Delta G_f^0 \text{ (products)} - \Sigma \Delta G_f^0 \text{ (reactants)} \tag{1.3}$$

so standard free energies of reactions may readily be computed from the tables. Converting to logarithms to the base ten at 25° and expressing energies in kcal/mole, (1.2) becomes,

$$-\Delta G^0 = 1 \cdot 36 \log K. \tag{1.4}$$

A large positive value of ΔG^0 for any reaction implies a very small equilibrium constant. The reactants are then thermodynamically stable with respect to the formation of the products because only very small amounts of the latter need be formed to reach equilibrium. Conversely, a large negative value of ΔG^0 implies a big equilibrium constant and instability of the reactants, because the formation of considerable quantities of products is necessary before equilibrium is achieved.

ΔG^0 is dependent only upon the initial and final states in a chemical equation. Thus thermodynamic stability or instability is independent of the path or mechanism of the reaction. The same cannot be said for kinetic stability. A chemical species is kinetically stable when it possesses one or more possible modes of decomposition which are thermodynamically favourable, but which proceed at immeasurably slow rates.

1.3. Kinetic stability. The rate equation for a chemical process,

$$a\text{A} + b\text{B} + c\text{C} + \ldots \rightarrow \text{products}, \tag{1.2}$$

usually takes the form,

$$\text{Rate} = k[\text{A}]^{n_a} [\text{B}]^{n_b} [\text{C}]^{n_c} \ldots, \tag{1.5}$$

---

[1] (1.2) is only valid when the equilibrium constant is expressed in the correct concentration units which depend upon the way in which the standard states are defined. With the definitions given in appendix 1, partial pressures in atmospheres should be used for a gas and molalities— the number of moles of solute in one thousand grams of solvent—for the solutes in a liquid solution.

where n_a is called the order of the reaction with respect to the reagent A, and k is the velocity constant. In more complex systems, the rate may be expressed as the sum of such terms. Experimentally, the variation of the velocity constant with temperature may usually be represented, to a good approximation, by the Arrhenius equation:

$$k = A\, e^{-E/RT}, \tag{1.6}$$

where A and E are positive quantities that change very little with temperature. As

$$\ln k = \ln A - E/RT, \tag{1.7}$$

a plot of $\ln k$ against $1/T$ gives a line of slope $-E/R$ from which E and A may be determined. Between them, (1.5) and (1.7) imply that the rate of chemical reaction is increased by heating. Denoting the parameters for the forward and backward reaction by the subscripts f and b respectively, (1.6) gives,

$$k_f/k_b = A_f/A_b\, e^{-(E_f - E_b)/RT}. \tag{1.8}$$

As

$$K = k_f/k_b \tag{1.9}$$

and, from (1.2),

$$K = e^{-\Delta G^0/RT} \tag{1.10}$$

$$= e^{+\Delta S^0/R} \cdot e^{-\Delta H^0/RT} \tag{1.11}$$

identification of the terms in (1.8) with those in (1.11) yields,

$$\Delta H^0 = E_f - E_b \tag{1.12}$$

and

$$A_f/A_b = e^{+\Delta S^0/R}. \tag{1.13}$$

These results suggest that the constant E should be interpreted as an enthalpy term. A way of doing this may be demonstrated by using the very simple process,

$$A + BC \to AB + C \tag{1.3}$$

as an example. A possible reaction mechanism involves the close approach of A to the molecule BC followed by expulsion of C with the rupture of the B–C bond. During this process, the energy of the system will increase from that of $(A + BC)$, reach a maximum with the formation of some configuration such as $(A \cdots B \cdots C)$, and

finally decline to that of (AB + C) with the elimination of C.
E may be equated to the enthalpy difference between (A $\cdots$ B $\cdots$ C),
which is called the activated complex, and (A + BC). It is then
known as the activation energy. These changes are illustrated by
the profile RQP in fig. 1.1. The reaction coordinate, which is plotted
along the x axis, is some parameter that represents the changes in

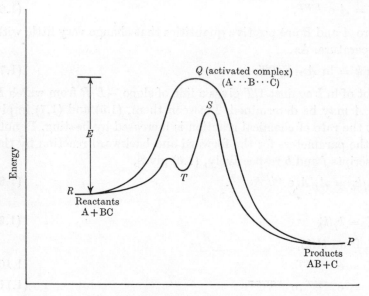

Fig. 1.1. Possible paths for a chemical reaction in the absence (RQP)
and presence ($RTSP$) of a catalyst.

molecular geometry that occur in the course of the reaction. The
probability of the reactants having an energy sufficient for the
formation of the activated complex will be low when the activation energy is high, and, for given concentrations, the process will
then be much slower. These facts are in qualitative agreement with
(1.5) and (1.6).[1]

The course of a reaction may be more complicated than it is for
this process. It may, for example, consist of a series of elementary
steps like RQP, and the energy profile will then show distinct
minima which may correspond to intermediates of an appreciable

---

[1] A more detailed treatment of the matters discussed in this section may
be found in standard texts on reaction kinetics, e.g. Laidler (1966).

lifetime. In this case, the energy of the highest transition state will determine the rate of the reaction.

The mechanism of a reaction is altered by the addition of a catalyst but the thermodynamics remain unchanged. In fig. 1.1, the catalyst might, for example, enable an intermediate T to be formed. If the equilibrium between T and the reactants was very rapidly established, the rate determining step would be the conversion of T to the products. This type of behaviour is common, but is only indirectly related to the accelerating role of the catalyst. The energy difference between the upper transition state S and the reactants is still the observed activation energy, and if this is less than E, then in the absence of a very unfavourable change in the pre-exponential factor, A, the catalysed reaction is faster. However, because the catalyst is unchanged when the reaction is over, the initial and final states remain the same. Thus the equilibrium is unperturbed.

1.4. The interpretation of stability. The single phenomenon of catalysis pinpoints a distinct difference between the sciences of kinetics and thermodynamics. In fig. 1.1, the speed of the reaction is much affected by the difference in the energy ordinates of points R and Q, while the position of equilibrium depends upon that of points R and P. Thus there can be no general connection between the rate of a reaction and its equilibrium constant, and no connection in specific cases unless a relation is established between the energy of reaction and the energy of activation. This point is rather difficult to accept, because there is undoubtedly some kind of overall inverse correlation between reaction velocity and standard free energy change. Indeed in some cases, such as the oxygen or hydrogen over-voltages that are required for the rapid decomposition of water, this correlation has become something of an empirical principle although exceptions are well-known (see §4.**10**). Nevertheless, immeasurably slow but very favourable reactions like the combustion of methane at room temperature:

$$CH_4 + 2O_2 \rightarrow CO_2 + 2H_2O, \quad \Delta G^0 = -196 \text{ kcal/mole} \qquad [1.4]$$

when contrasted with the speed of processes which have almost zero free energy change:

$$Fe(H_2O)_6^{2+} + D_2O \rightarrow Fe(H_2O)_5(HDO)^{2+} + HDO \qquad [1.5]$$

enforce general acquiescence in the more pessimistic view. Consequently a compound may be stable either for kinetic or for thermodynamic reasons, and no conclusion as to the one may be drawn from the other. If therefore an explanation is advanced for the stability of a compound, it is essential to decide whether this explanation applies to the rate or to the equilibrium constant of the reaction, for until this analysis has been made, the interpretation cannot be complete.

The distinction between thermodynamic and kinetic stability should always be remembered. Neither the theory of the kinetics or of the thermodynamics of reactions could be said to be at an advanced stage, but because a single reaction may proceed by a number of mechanisms, and because their short lifetimes make the characterization and investigation of transition states very difficult, the theory of kinetics is less well understood. This book is concerned almost entirely with thermodynamic stability.

It is an instructive exercise to examine the limitations that this approach imposes upon an assessment of the stability of a compound. In pursuing this question, the standard free energy of formation of the compound would first have to be obtained, either by calculation or by experimental measurement. From the standard free energies of formation of all possible decomposition products, the standard free energies of all decomposition reactions could then be determined. If all these values of ΔG^0 were positive, the compound would be stable. If one or more were negative, it might be stable and it might not, for a thermodynamically favourable reaction could be inhibited for kinetic reasons. Clearly the interpretation of chemical stability or instability is a very formidable task, and thermodynamics alone yield only a partial solution. Nevertheless, compared with the kinetic approach it is more important, for a reaction with a positive standard free energy change will never proceed to virtual completion, while if the standard free energy change is negative, a very slow reaction may sometimes be speeded up by catalysis or an increase in temperature.

When discussing the instability of a compound, it is essential to consider the reaction to which it is unstable. This apparently obvious point is sometimes ignored. For example, an interpretation of the negative standard free energy of formation of a compound is sometimes cited as an explanation of its stability. There is, how-ever, little doubt that the compounds $ScCl_2$ and $MnCl_3$ which are unknown at 25° have very negative values of ΔG_f^0, but their instability is related to the decomposition reactions

$$3ScCl_2 \to Sc + 2ScCl_3, \qquad\qquad [1.6]$$

$$MnCl_3 \to MnCl_2 + \tfrac{1}{2}Cl_2, \qquad\qquad [1.7]$$

which have negative standard free energies at this temperature. Again, theoretical chemistry is much concerned with the stabilities of compounds with respect to their constituent atoms which in nearly all cases is only indirectly related to real thermodynamic stability. Where the word 'stability' is used or implied in this book, an attempt has been made to relate it to a particular reaction or decomposition.

One more difficulty in the interpretation of stability should be mentioned. (1.4) shows that, at 25°, a factor of ten in the equilibrium constant is equivalent to only 1·36 kcal/mole. Its logarithmic dependence means that a calculated equilibrium constant is very sensitive to small errors in ΔG^0. As any kind of theoretical approach to the estimation of equilibrium constants is usually based on the calculation of energies, (1.4) quite plainly implies that the interpretation of stabilities is a very formidable problem.

1.5. Enthalpy and entropy terms.
When considering the free energies of reactions, it is often convenient to split them into two parts. For any isothermal process:

$$\Delta G = \Delta H - T\Delta S, \qquad\qquad (1.14)$$

where ΔS is the entropy change and ΔH is the enthalpy change, which is related to q_p, the heat absorbed at constant pressure, by the equation,

$$\Delta H = q_p. \qquad\qquad (1.15)$$

The standard enthalpy change of a reaction occurring at constant temperature may be calculated if the standard enthalpies of formation of the reactants and products are known, for then

$$\Delta H^0 = \Sigma \Delta H_f^0 \text{ (products)} - \Sigma \Delta H_f^0 \text{ (reactants)}. \qquad\qquad (1.16)$$

Because the standard molal entropies of elements or compounds may be determined by using the third law of thermodynamics, it is these, rather than standard entropies of formation, that are recorded in most compilations. For any reaction,

$$\Delta S^0 = \Sigma S^0 \text{ (products)} - \Sigma S^0 \text{ (reactants)} \tag{1.17}$$

The melting of a solid or the boiling of a liquid occurs reversibly at the transition temperature. Thus,

$$dS = q/T \tag{1.18}$$

and

$$\Delta S = L/T, \tag{1.19}$$

where L is the enthalpy of fusion or evaporation, and T is the melting or boiling point. As latent heats of evaporation usually exceed latent heats of fusion by a substantial amount, the entropies of gases tend, in general, to be considerably larger than those of liquids, which in turn are greater than those of solids. Some sample figures are presented in table 1.1, and more examples can be found in the tables in the appendix.

TABLE 1.1 *Standard molal entropies of some substances at 25° in cal/deg.mole*

Substance	S^0	Substance	S^0
H_2 (g)	31·2	NF_3 (g)	62·3
F_2 (g)	48·4	BF_3 (g)	60·7
Br_2 (l)	36·4	BCl_3 (g)	69·3
Br_2 (g)	58·6	BBr_3 (g)	77·5
I_2 (s)	27·8	ClO_2 (g)	61·4
I_2 (g)	62·3	SO_2 (g)	59·3
Graphite	1·4	OsO_4 (s)	32·2
C (g)	37·8	OsO_4 (g)	70·1
HCl (g)	44·6	MoF_6 (g)	80·2
NH_3 (g)	46·0	Al_2Cl_6 (g)	117
NH_4Cl(s)	22·6	NO_2 (g)	57·4
H_2O(l)	16·7	N_2O_4 (g)	72·7
H_2O (g)	45·1	$TiCl_4$ (l)	59·5
		$TiCl_4$ (g)	84·3

The entropies of gases increase with molecular weight and molecular complexity. According to statistical mechanics, the separations of the quantized energy levels for translational motion diminish with particle mass, and those for rotational motion about an axis fall as the moment of inertia is increased. Consequently, at a given temperature, more energy levels are occupied in heavy, complex molecules, and there are a correspondingly larger number of ways in which the individual molecules in one mole may be assigned to them. This necessarily implies that the molal entropy is greater. Because the property is mainly dependent on mass and molecular geometry, the unknown entropy of a gas may sometimes be estimated quite accurately by using the value for a carefully chosen analogue. Compare, for example, the values for ClO_2 and SO_2 in table 1.1.

The entropies of solids may be estimated by means of empirical rules proposed by Latimer (1952). He pointed out that the standard molal entropy of a solid compound could be represented quite accurately as the sum of constants characteristic of the elements from which it is composed. Initially he proposed the formula,

$$S = \tfrac{3}{2}R \ln M - 0 \cdot 94, \tag{1.20}$$

where S is the constant for any element and M is its atomic weight. Later the values were adjusted slightly to give a better empirical fit. These figures are presented in table 1.2. Figures in brackets are placed against those elements whose contributions vary slightly with the oxidation number of the metal with which they are combined. In the reference quoted, Latimer also gives values for the calculation of the entropies of oxides and compounds containing complex anions. In estimating the entropy of a solid, it is usually better to choose an analogous compound with a known entropy, and estimate the difference between the known and unknown values from table 1.2, rather than compute S^0 directly from the individual contributions of the elements. Thus $S^0[Na_2SO_4] = 37 \cdot 4$ cal/deg.mole. Adding twice the difference between the sodium and silver contributions in table 1.2, we obtain an estimated $S^0[Ag_2SO_4]$ of $48 \cdot 0$ cal/deg.mole compared with an experimental value of $47 \cdot 9$ cal/deg.mole.

As the entropies of vapours are usually rather bigger than those of solids or liquids, the entropy change assumes a greater numerical significance when there is a change in the number of moles of gas in the course of a reaction. *Under certain conditions*, a large entropy change has a profound effect on the equilibrium position of a reaction when the temperature is increased. On heating, the

TABLE 1.2[a]   *Entropy contributions of the elements in solid compounds at 25° in cal/deg.mole*

Element	S^0	Element	S^0	Element	S^0	Element	S^0
Ag	12·8	Dy	14·4	Mn	10·3	Se	(11·6)
Al	8·0	Er	14·5	Mo	12·3	Si	8·1
As	11·45	Eu	14·1	N	5·8	Sm	14·1
Au	15·3	F	(6·9)	Na	7·5	Sn	13·1
B	4·9	Fe	10·4	Nd	13·9	Sr	12·0
Ba	13·7	Ga	11·2	Ni	10·5	Ta	14·9
Be	4·3	Gd	14·3	Os	15·1	Tb	14·3
Bi	15·6	Ge	11·3	Pb	15·5	Te	(13·4)
Br	(11·7)	Hf	14·8	Pd	12·7	Th	15·9
C	5·2	Hg	15·4	Pr	13·8	Ti	9·8
Ca	9·3	Ho	14·5	Pt	15·2	Tl	15·4
Cb	12·2	I	(13·4)	Ra	15·8	Tm	14·6
Cd	12·9	In	13·0	Rb	11·9	U	16·0
Ce	13·8	Ir	15·2	Re	15·0	V	10·1
Cl	(8·8)	K	9·2	Rh	12·5	W	15·0
Co	10·6	La	13·8	Ru	12·5	Y	12·0
Cr	10·2	Li	3·5	S	(8·5)	Yb	14·7
Cs	13·6	Lu	14·8	Sb	13·2	Zn	10·9
Cu	10·8	Mg	7·6	Sc	9·7	Zr	12·1

[a] Reprinted from Latimer (1952). *The Oxidation States of the Elements and their Potentials in Aqueous Solution.* 2nd edition. New Jersey: Prentice-Hall.

equilibrium is displaced in a direction that is determined by the sign of ΔH^0, in accordance with the van't Hoff isochore:

$$\frac{d \ln K}{dT} = \frac{\Delta H^0}{RT^2}. \tag{1.21}$$

If ΔG^0 is positive or only slightly negative, a large positive value of ΔS^0 will ensure that the reaction is appreciably endothermic at 25°. Thus if a process is unfavourable at room temperature but has a large positive entropy change, it may well proceed when the system is heated. In the reactions,

$$N_2O_4(g) \rightarrow 2NO_2(g), \quad \Delta G^0 = 1\cdot 1 \text{ kcal/mole},$$
$$\Delta S^0 = 42\cdot 0 \text{ cal/deg.mole} \quad \text{[1.8]}$$

and

$$NH_4Cl(s) \rightarrow NH_3(g) + HCl(g), \quad \Delta G^0 = 21\cdot 8 \text{ kcal/mole},$$
$$\Delta S^0 = 68\cdot 0 \text{ cal/deg.mole}, \quad \text{[1.9]}$$

the positive values of ΔS^0 associated with the increases in the number of moles of gas, taken in conjunction with the sign of ΔG^0, ensure decomposition when the temperature is raised. An example of the reverse effect is supplied by the reaction,

$$2CO\,(g) \rightarrow C\,(s) + CO_2\,(g), \quad \Delta G^0 = -28{\cdot}7 \text{ kcal/mole},$$
$$\Delta S^0 = -42{\cdot}0 \text{ cal/deg.mole.} \qquad [\textbf{1.10}]$$

Carbon monoxide is the predominant constituent of the equilibrium mixture above about 700°.

TABLE 1.3 *Entropies of decomposition for the carbonates of some metals in cal/deg.mole*

Carbonate	ΔS^0	Carbonate	ΔS^0
$BaCO_3$	41·1	$MgCO_3$	42·0
$CdCO_3$	39·0	$MnCO_3$	44·9
$CaCO_3$	39·4	$SrCO_3$	40·9
$CuCO_3$	40·3	$ZnCO_3$	41·9

At normal temperatures, the entropy term is in general less effective than ΔH^0 in determining the sign of the free energy change. At 25° for example, three cal/deg.mole are equivalent to less than one kcal/mole. In comparative inorganic chemistry, the significance of the entropy change is often even less. The figures in table 1.3 show that the standard entropy of the decomposition of certain metal carbonates:

$$MCO_3\,(s) \rightarrow MO\,(s) + CO_2\,(g) \qquad [\textbf{1.11}]$$

is almost independent of the metal. This observation is a necessary corollary of Latimer's rule, for the individual contributions of the metal to the entropies of the solid carbonate and solid oxide are the same.

A similar result is obtained when the entropies for the gas phase decomposition of the boron halides are evaluated (see table 1.4). The rise in the entropy of the halide with increasing mass and moment of inertia of the molecule is roughly matched by that of the free halogen. Thus in both these cases, the variations in the standard free energy of decomposition are almost entirely determined by those in the corresponding enthalpy change.

TABLE 1.4   *Standard molal entropies of the reaction:*

$$BX_3(g) = B(s) + \tfrac{3}{2}X_2(g) \ \text{at } 25° \text{ in cal/deg.mole}$$

BF$_3$	BCl$_3$	BBr$_3$	BI$_3$
13·4	12·0	11·9	11·4

Summarizing, large entropy changes are often encountered when there is a rise or fall in the number of moles of gas in the course of a reaction. The solution chemistry of electrolytes is another field where entropy changes are often both important and substantial (see §5.2). However, of the two contributions to the free energy of a reaction, the entropy term is in general both less significant and more susceptible to theoretical interpretation than the enthalpy change. This is particularly the case in comparative chemistry. The variations in the enthalpies of reactions are much more difficult to understand or predict.

1.6.   The role of thermodynamics in interpretative chemistry.   It is very important to appreciate that thermodynamics cannot, by itself, supply explanations for chemical phenomena. A precise and self-consistent role for thermodynamics in the interpretation of chemistry (Nelson and Sharpe, 1966) may be illustrated by reference to a specific problem; namely that, while gaseous nitrogen trifluoride is thermodynamically stable with respect to its elements at 25°, liquid nitrogen trichloride explodes spontaneously to form nitrogen and chlorine at the slightest provocation.

If this problem is to be explored at all, it must first be stated in numerical terms. This is easily done; why is ΔG_f^0 [NF$_3$(g)] negative while ΔG_f^0 [NCl$_3$(g)] is positive? The difference between the two standard free energies of formation is in fact some 80–90 kcal/mole, so although NCl$_3$ is a liquid at 25°, we may neglect its small free energy of vaporization and consider the stability of the gas.

The second stage of the problem also involves thermodynamics: the problem stated in the first stage is restated by means of a thermodynamic cycle. One possible cycle[1] is that shown in fig. 1.2. As the values of ΔS_f^0 are very similar for both compounds, this need only be expressed in terms of enthalpies. From fig. 1.2,

$$\Delta G_f^0 \,[NCl_3(g)] - \Delta G_f^0 \,[NF_3(g)]$$

$$\simeq \Delta H_f^0 \,[NCl_3(g)] - \Delta H_f^0 \,[NF_3(g)] \quad (1.22)$$

$$= \tfrac{3}{2}(D[Cl_2] - D[F_2]) - 3(B[N\text{–}Cl] - B[N\text{–}F]). \quad (1.23)$$

The term in the first bracket of (1.23) is positive, while that in the second is negative. Thus the conclusion of the restatement implied by fig. 1.2 is that the greater stability of NF_3 is due to the lower dissociation energy of fluorine and the stronger bonds that the fluorine atoms form with nitrogen.

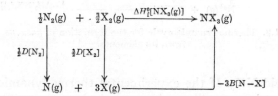

Fig. 1.2. Thermodynamic cycle for the formation of gaseous NX_3 from its elements.

Fig. 1.2, however, does not contain the only restatement of the original problem. Another is shown in fig. 1.3, where $\Delta H_I[NX_3(g)]$ is the standard enthalpy change associated with the formation of gaseous N^{3+} and X^- ions from the gaseous halide. Here,

$$\Delta H_f^0 \,[NCl_3(g)] - \Delta H_f^0 \,[NF_3(g)] = 3(\Delta H_f^0 \,[Cl^-(g)] - \Delta H_f^0 \,[F^-(g)])$$

$$- (\Delta H_I \,[NCl_3(g)] - \Delta H_I \,[NF_3(g)]). \quad (1.24)$$

[1] In all thermodynamic cycles shown in this book, the symbols represent enthalpy or free energy changes for the direction pointed out by the arrow. Definitions of the symbols used, and the quantities that they represent may be found in appendix 1. In moving clockwise (or anti-clockwise) round the cycle from A to B, the overall enthalpy or free energy change may then be computed by adding the symbols attached to the intervening clockwise (anti-clockwise) arrows and subtracting those attached to the anti-clockwise (clockwise) arrows. The overall clockwise and anti-clockwise values obtained in this way are equal by Hess's Law.

From table 2.10, the values of $\Delta H_f^0 [X^- (g)]$ are quite similar for fluorine and chlorine, so the new restatement is that the greater stability of NF_3 is due to the greater energy released when the small gaseous N^{3+} and F^- ions combine to form the NF_3 molecule.

It is apparent that there are an infinite number of cycles which might be used to restate the original problem. We might, for example, consider the formation of N^{6+} and X^{2-} ions, or of N^{3-} and X^+ ions from nitrogen and the halogen, followed by their combina-

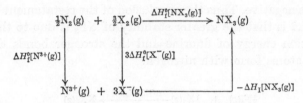

Fig. 1.3. Thermodynamic cycle for the formation of gaseous NX_3 from its elements.

tion to yield NX_3. If the experimental thermodynamic data are available, each cycle will provide a conclusion concerning the interplay of the terms within it, and their relative importance in determining the thermodynamic stabilities of NF_3 and NCl_3 with respect to their elements. Each conclusion will be as quantitatively precise as the experimental information used in its derivation, and every one will be equally valid. Nevertheless, however illuminating it may be, each conclusion is no more than a restatement of the original problem, the only change being that it demands an explanation of the relative values of a different thermodynamic term or terms from that considered in the initial statement.

At this stage it is necessary to select one of the restatements for further investigation. The choice of restatement will be determined by two things: first, the precision of the experimental values of the thermodynamic quantities in the cycle on which the restatement is based. Thus the cycle involving N^{6+} and X^{2-} ions might be rejected because the standard enthalpies of the reactions

$$\tfrac{1}{2}X_2(g) + 2e \rightarrow X^{2-}(g) \qquad\qquad\qquad [1.12]$$

are unknown. The second but most important consideration concerns our ability to interpret, by means of theoretical chemistry, the values of the thermodynamic quantities in the cycles. For example, in deciding between the first and second restatements, we would consider whether current theories were more capable of explaining the higher energy evolved in the recombination of gaseous N^{3+} and F^- ions, than they were of interpreting the lower dissociation energy of fluorine and the stronger bonds that this halogen forms with nitrogen.

Once this choice has been made, we enter upon the third stage of the investigation, the theoretical interpretation of the restated problem. Here we leave the realm of thermodynamics and enter that of theoretical chemistry, for thermodynamics only states and restates the problem, it does not pretend to offer a solution. It will be noted that no one restatement is more 'correct' than another. Thus the first restatement is not to be preferred to the second because the nitrogen halides are covalent compounds. Of the two cycles, it is difficult to say which is the more suitable, because current theories are no more capable of giving a quantitative interpretation of the strengths of the N—F and N—Cl bonds than they are of the energies of recombination of ions to give NF_3 and NCl_3. If the two halides were ionic solids, the second type of cycle would probably be chosen because quantitative theoretical interpretations of the lattice energies of crystals are sometimes possible.

Summarizing then, an investigation of a chemical problem by thermodynamic methods can be divided into three parts. In the first part, the problem is stated by means of thermodynamics. Thermodynamics is then used again in the second part to restate the problem in the one of many possible ways that is the most susceptible to theoretical study. Once this has been done, the job of thermodynamics is finished and the third and final part of the investigation, the theoretical interpretation of the restated problem can begin. Thus thermodynamics acts as a bridge between theoretical chemistry and the questions posed by experiment. If the theoretical response is sometimes only qualitative or unconvincing, this failing cannot be attributed to thermodynamics. Indeed, such failures are often a reflection of the value of this approach, for through its rigorous nature, thermodynamics makes very exacting demands of theoretical chemistry. It does not, for example, ask why gaseous nitrogen trifluoride is stable with respect to its elements in their standard states, but why the value of

ΔG_f^0 [NF$_3$(g)] at 25° is -20.0 ± 3 kcal/mole. The accuracy of the response made to this question by a particular theory is an invaluable measure of the success of that theory.

1.7. Thermodynamic correlations. When a problem is investigated by the method described in the previous section, the inadequacy of current theories often prevents any useful result being obtained at the third stage. Nevertheless the restatements for a number of different problems are sometimes very similar, and this constitutes a useful correlation. In chapter 6, for example, the variations in the relative stabilities of the dipositive and tripositive oxidation states of the lanthanides and the metals of the first transition series and of the tripositive and tetrapositive oxidation states of the actinides in certain environments are correlated with a saw-tooth variation in the appropriate ionization potential. Similarly, among the gaseous molecular halides, many different fluorides are more stable with respect to their elements than the corresponding chloride, bromide or iodide. In each case, this may be attributed to the low dissociation energy of fluorine and the stronger bonds formed by the free atom. Even if the possibility of interpretations of these thermodynamic quantities is excluded, such correlations are very valuable.

References

Laidler, K. J. (1966). *Chemical kinetics*, 2nd edition. New York: McGraw-Hill.

Latimer, W. M. (1952). *The Oxidation States of the Elements and their Potentials in Aqueous Solution*, 2nd edition. New Jersey: Prentice Hall.

Nelson, P. G. and Sharpe, A. G. (1966). *J. Chem. Soc.* (A), 501.

Rossini, F. D. (1964). *Pure and Applied Chem.* **9**, 453.

2. The Ionic Model

2.1. Introduction. Because of similarities in their properties, certain simple chemical compounds may be grouped together in a particular class. They are solid and involatile, and although their electrical conductivity is very low in the solid state, it is markedly increased by fusion. Several binary compounds of this type have been subjected to a very careful examination by X-ray techniques. This showed that the electron density distribution falls to a very low value between adjacent unlike nuclei, and that if the distribution is divided at this point, each atom possesses an approximately integral net charge. In sodium chloride and calcium fluoride for example, electrons are transferred from chlorine to sodium and from fluorine to calcium to give the charges $Na^+ Cl^-$ and $Ca^{2+} 2F^-$ on the constituent atoms. Although the X-ray analyses are not accurate enough to show how complete the electron transfer is, the results suggest that, to a first approximation, we may view these substances as composed of ions.

A further discovery made by using X-ray techniques, was that the binary compounds of this type possessed lattices in which the atoms were regularly disposed. Thus in sodium chloride, each sodium is surrounded by six chlorines and vice versa, suggesting that this arrangement is preferable to the juxtaposition of similarly charged ions. The structures of sodium chloride and calcium fluoride are shown in figs. 2.1 and 2.2 (reprinted from Wells, 1962). In the fluorite lattice there is eight coordination of calcium by fluorine, and four coordination of fluorine by calcium. The same experiments disclosed that the internuclear distances in the compounds were approximately additive. For example, in table 2.1 the differences in the nuclear separations of the rubidium and potassium, potassium and sodium, and sodium and lithium halides are shown, and they are nearly independent of the halide ion concerned.

All these observations are compatible with a model that regards

[17]

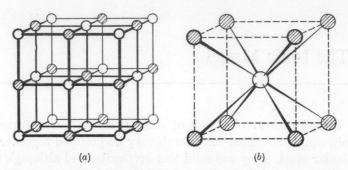

Fig. 2.1. The sodium chloride (*a*) and caesium chloride (*b*) structures.

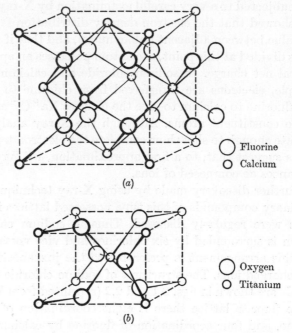

Fluorine
Calcium

(*a*)

Oxygen
Titanium

(*b*)

Fig. 2.2. (*a*) The fluorite (CaF_2) and (*b*) rutile (TiO_2) structures.

these compounds as composed of integrally charged spheres when the electron distribution of the constituent ions has this symmetry. These spheres are usually viewed as fairly rather than perfectly hard because, as table 2.1 shows, the additivity of the internuclear distances is not perfect and because the contributions

TABLE 2.1   *Additivity of internuclear distances in alkali metal halides with the rock salt structure*

	Li	Na	K	Rb	Cs
$r_0(MCl)$-$r_0(MF)$	0·56	0·50	0·47	0·47	0·46
$r_0(MBr)$-$r_0(MCl)$	0·18	0·17	0·15	0·14	
$r_0(MI)$-$r_0(MBr)$	0·25	0·25	0·23	0·24	
	F	Cl	Br	I	
$r_0(NaX)$-$r_0(LiX)$	0·30	0·25	0·24	0·23	
$r_0(KX)$-$r_0(NaX)$	0·36	0·33	0·31	0·30	
$r_0(RbX)$-$r_0(KX)$	0·15	0·14	0·13	0·14	

of individual ions to the internuclear distance increases with their coordination number. Thus the normal eight coordinate structure of caesium chloride (see fig. 2.1) can be transformed under pressure to the six coordinate rock salt form with a reduction of about 0·10 Å in the interatomic distance. It should be noted that the precise mathematical nature of the forces opposing compression is not usually defined.

A finite sphere picture is not compatible with wave mechanics which implies that the electron density distribution around any ion has a finite value even at very large distances from the nucleus. Nevertheless, if it is used as a model, it is possible to assign radii to individual ions and to discuss, with limited success, the variation of crystal structures with ionic size. Unfortunately the ionic model cannot be defined in this simple way because the present conception of it is a combination of ideas developed for energetic as well as crystallographic purposes. As we shall see, a compound is often regarded as ionic for energetic reasons if its experimental lattice energy can be accurately predicted by a semi-theoretical expression. Although these two separate aspects of the ionic model make its precise definition a difficult task, they are not entirely unconnected. The hard sphere picture, which is crystallographically very useful, points the way towards suitable expressions for the lattice energy.

2.2. The lattice energy.

The lattice energy of a compound, U_0, at 0 °K, may be defined as the internal energy change observed when one mole of the compound at one atmosphere pressure is converted into defined gaseous ions which are infinitely removed from one another. At 0 °K these ions are stationary.

The lattice energy at 298 °K, U_{298}, is related to other important quantities by the Born–Haber cycle which, for the compound MX_n, is shown in fig. 2.3.

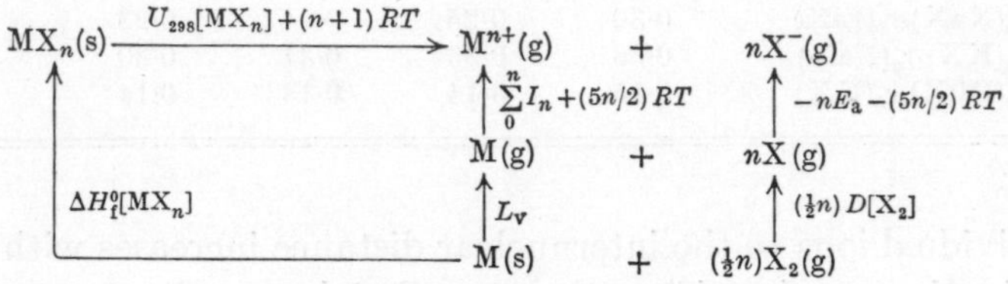

$$MX_n(s) \xrightarrow{\;U_{298}[MX_n]+(n+1)RT\;} M^{n+}(g) \quad + \quad nX^-(g)$$

with vertical relations:

$\uparrow \sum_0^n I_n + (5n/2)RT$ $\uparrow -nE_a - (5n/2)RT$

$$M(g) \quad + \quad nX(g)$$

$\Delta H_f^0[MX_n]$ $\uparrow L_v$ $\uparrow (\tfrac{1}{2}n)D[X_2]$

$$M(s) \quad + \quad (\tfrac{1}{2}n)X_2(g)$$

Fig. 2.3. The Born–Haber cycle.

In the cycle, L_v is the latent heat of sublimation of the metal, $\sum_0^n I_n$ is the sum of the first n ionization potentials, $E_a[X]$ is the electron affinity of X and $D[X_2]$ is the dissociation energy of X_2 at 298 °K. These quantities are defined in appendix 1. It is assumed that the standard state of X at 298 °K is $X_2(g)$, and that the excited states of the gaseous atoms and ions involved in the cycle are not appreciably populated at this temperature. The ions $M^{n+}(g)$ and $X^-(g)$ in fig. 2.3 are in the hypothetical ideal gas state of one atmosphere pressure, so one does not need to consider any interactions between them. The derivation of the standard enthalpies in the cycle from the quantities defined above is given in appendix 2.

From fig. 2.3,

$$U_{298}[MX_n] = L_v + \sum_0^n I_n + \frac{n}{2}D[X_2] - nE_a[X]$$
$$- \Delta H_f^0[MX_n] - (n+1)RT. \quad (2.1)$$

If, for the standard enthalpy of the process,

$$M(s) \rightarrow M^{n+}(g) + ne^-(g) \qquad\qquad\qquad\qquad [2.1]$$

we write $\Delta H_f^0[M^{n+}(g)]$, and for that of the reaction,

$$\frac{n}{2}X_2 \text{ (ref. state)} + ne^-(g) \rightarrow nX^-(g), \qquad\qquad\qquad [2.2]$$

$n\Delta H_f^0[X^-(g)]$, (2.1) becomes,

$$U_{298}[MX_n] = \Delta H_f^0[M^{n+}(g)] + n\Delta H_f^0[X^-(g)]$$
$$- \Delta H_f^0[MX_n] - (n+1)RT. \quad (2.2)$$

The lattice energy at 298 °K may be corrected to 0 °K by the method described in appendix 2. The difference between U_{298} and U_0 is not more than 3 kcal/mole for the halides of the alkali metals or the alkaline earths. For these compounds, independent methods now exist by which all the quantities on the right-hand sides of (2.1) and (A2.12) (in appendix 2) have been determined. The resulting data can be used to calculate values of U_0 and figures derived for some of the halides are shown in the third column of table 2.2. The uncertainties in these figures are ± 3–4 kcal/mole.

If the compounds were composed of fairly hard spheres and the charges on the anion and cation were z_1 and z_-, then the coulombic

TABLE 2.2 *Lattice energies of alkali metal and alkaline earth dihalides at 0 °K (kcal/mole)*

Compound	Structure	Born cycle[a]	Equation (2.7)[b]	Extended calculation[c]
LiF	Rock salt	244		247
LiI	Rock salt	181		177
NaF	Rock salt	216	216	219
NaCl	Rock salt	185	181	186
NaBr	Rock salt	176	172	177
NaI	Rock salt	168	161	165
KCl	Rock salt	168	165	169
KI	Rock salt	155	149	153
CsF	Rock salt	175	173	179
CsCl	CsCl	157	149	156
CsI	CsCl	146	136	144
MgF$_2$	Rutile	695	689	696
CaF$_2$	Fluorite	619	620	624
CaCl$_2$	Deformed Rutile	531		531
SrCl$_2$	Fluorite	508	494	509
BaF$_2$	Fluorite	553	559	560
BaBr$_2$	PbCl$_2$	465		466
BaI$_2$	PbCl$_2$	447		438

[a] Calculated from the sources mentioned in appendix 1. ΔH_f^0 values are from NBS Circular 500 unless otherwise recommended by Pitzer and Brewer (1961).
[b] Morris (1957).
[c] Cubiociotti (1961); Brackett and Brackett (1965).

energy of interaction between a pair of oppositely charged ions
would be $-z_+z_-e^2/r$, where r is the internuclear distance. In a real
crystal, any one ion interacts with a very large number of others,
and the computation of the total coulombic energy involves the
summation of an infinite series. In the case of the rock salt lattice
in fig. 2.1, the central sodium ion interacts first with its six neigh-
bouring chlorines, then with the twelve sodiums at the mid-points
of the edges, then with the eight chlorines at the corners, then with
the six sodiums at the centres of the six adjacent unit cells, and so
on. The coulombic energy of interaction is therefore given by the
expression,

$$-e^2/r(6/\sqrt{1} - 12/\sqrt{2} + 8/\sqrt{3} - 6/\sqrt{4} + ...), \tag{2.3}$$

where r is the internuclear distance. The term in the bracket is
known as the Madelung constant, M. The summation is difficult
because of its very slow convergence, but values have been obtained for a number of the simpler crystal structures. The final expression obtained for the coulombic energy per mole is $-N_0 M z_+z_-e^2/r$,
N_0 being the Avogadro number. Some values of the Madelung
constant which are appropriate when it is defined by this formula
for the coulombic term are given in the second column of table 2.8.

Viewing the ions as fairly hard spheres, the coulombic interaction
is opposed by a repulsive force which, once the ions come into
contact, increases very rapidly as the internuclear distance is
shortened. A possible function with this characteristic is B/r^n
where n is large and both it and B are constants. A trial expression
for the lattice energy of a crystal at 0 °K is therefore:

$$U_0 = \frac{N_0 M z_+z_-e^2}{r} - \frac{B}{r^n}, \tag{2.4}$$

$$\frac{dU_0}{dr} = -\frac{N_0 M z_+z_-e^2}{r^2} + \frac{nB}{r^{n+1}}. \tag{2.5}$$

As $dU_0/dr = 0$ at the equilibrium internuclear distance, r_0,

$$B = \frac{N_0 M z_+z_-e^2 r_0^{n-1}}{n}, \tag{2.6}$$

$$\therefore U_0 = \frac{N_0 M z_+z_-e^2}{r_0}\left(1 - \frac{1}{n}\right), \tag{2.7}$$

(2.7) is known as the Born–Landé equation. r_0 may be determined by X-ray crystallography, and n can be calculated from compressibility measurements by using the relation derived in appendix 3. It usually lies between 5 and 12.

Some lattice energies obtained from (2.7) for some alkali metal and alkaline earth dihalides are shown in the fourth column of table 2.2. The agreement between the Born–Haber cycle lattice energies and the compressibility/r_0 values is good, although there are noticeable deviations with compounds containing the larger anions. More recent workers prefer the expression $a\mathrm{e}^{-r/\rho}$ for the repulsive force where a and ρ are constants for a particular compound, again determined by compressibility measurements. This yields the Born–Mayer relation:

$$U_0 = +\frac{N_0 M z_+ z_- e^2}{r_0}\,(1-\rho/r_0).\tag{2.8}$$

The disagreement between the formulae and experimental values can be lowered by allowing for the London or van der Waals forces which are attributed to the synchronization of the oscillations of the electrons on the ions, and for the zero-point energy, but the process is a highly empirical one involving the use of several experimental parameters. Some values obtained by this method of 'extended classical calculation' are shown in column five of table 2.2.

Equations (2.7) and (2.8) cannot strictly be described as 'theoretical' because experimental values of r_0 and n or ρ are required. Although agreement between compressibility and cycle lattice energies is suggestive, it does not constitute a proof of the validity of a hard sphere or any other model which sets up an ionic charge distribution. (2.7) and (2.8) embody self-compensating features which tend to promote such agreement (see §2.4).

2.3. Further tests of the compressibility equations.
Because of the empiricism of the expressions introduced in the previous section, it is important to check them more widely. Such a check is only possible for salts of anions X^{n-} when the quantity $\Delta H_f^0[X^{n-}(g)]$ is known from experimental measurements which are independent of the Born–Haber cycle. Such measurements are difficult, and apart from the halide figures, the only values with a

reliability sufficient for them to be quoted by the Revisers of the National Bureau of Standards Circular 500 are those for the ions H^-, OH^- and N_3^-. Of these three values, the figure for the azide ion probably has a rather large uncertainty because it was determined by electron impact methods. In experiments of this kind, it is difficult to allow for the kinetic energy of the products of the impact.

TABLE 2.3 *Lattice energies of some azides, hydrides and hydroxides at 0 °K (kcal/mole)*

Compound	Born cycle[a]	Compressibility[b] equations
KN_3	164	156
RbN_3	158	149
CsN_3	153	142
NaOH	213	197
KOH	190	175
LiH	218	222
NaH	192	191
KH	170	169
RbH	161	161
CsH	152	155

[a] ΔH_f^0 for azides: Gray and Waddington (1956). ΔH_f^0 for hydrides: Gibb (1962).
[b] Gibb (1962); Gray and Waddington, (1956) and Waddington (1959).

The hydrides of the alkali metals have the rock salt lattice, while above 200°, NaOH and KOH also adopt this structure because the hydroxide ions in what were the low temperature orthorhombic forms become freely rotating. The linearity of the azide anion in KN_3, RbN_3 and CsN_3 leads to tetragonal structures, but the Madelung constants can still be calculated, and lattice energies obtained in this way are presented in table 2.3. The extended calculation formula used for the azides included an additional term not previously mentioned here which allows for the influence of the multipole produced by the non-spherical symmetry of the anion. For the hydrides the agreement with the cycle values is excellent, and although the differences for the azides and hydroxides are sufficient to instil caution, the compressibility data was not very reliable and this, together with possible errors in the values of $\Delta H_f^0[X^-(g)]$, could well account for any discrepancies.

All of the compounds in tables 2.2 and 2.3 possess the characteristics described in §2.1, and when this is the case and the data used are reliable, it seems that the agreement between cycle values and lattice energies calculated by the methods of the previous section is good.

2.4. The lattice energy as a criterion of bond type.

To define what is meant by an ionic compound is not a simple task. Ideally, a substance should be called 'ionic' if an analysis of the electron density distribution shows that, within small well defined limits, each atom or complex radical bounded by distribution minima carries an integral net charge. Unfortunately such analyses have been made in relatively few cases, and they are not yet numerous or accurate enough to make this a practical proposition. The results of §2.2 and §2.3 suggest an alternative although less direct approach. This is to define an ionic compound as one whose lattice energy can be reproduced, within the limits of experimental error, by the methods of calculation described in those sections. This definition is not directly concerned with the electron density distribution along the cation–anion axis, a fact that is taken up in more detail later in the section.

It is interesting to see whether the definition is consistent with those other properties, mentioned in §2.1, which are usually taken as evidence of the presence of ions in crystals. These properties are:

(a) Existence as an involatile[1] solid at normal temperatures.

(b) Poor conductance in the solid state but good when fused.

(c) Possession of a three-dimensional lattice in which the atoms are regularly disposed. In this lattice, the nearest neighbours of any ion in any direction are ions of opposite charge.

A definition via the lattice energy is selective in that it sometimes eliminates compounds, such as HgF_2 and AgI, which display these three characteristics. As tables 2.4 and 2.5 show, the calculated lattice energies of these salts differ from their cycle values by 36 and 27 kcal/mole respectively. To demonstrate consistency however, it is sufficient to show, first that compounds which con-

---

[1] In compounds containing complex ions, interionic reactions may occasionally cause low volatility. X-ray studies of solid phosphorous pentachloride suggest the formulation $PCl_4^+PCl_6^-$, but the compound sublimes to give PCl_5 on mild heating.

TABLE 2.4   *Lattice energies of the silver halides at* 0 °K (*kcal/mole*)

Compound	Born cycle[a]	Extended calculation[b]
AgF	228	221
AgCl	216	199
AgBr	212	193
AgI	212	185

[a] See footnote (a) to table 2.2.   [b] Ladd and Lee (1964).

TABLE 2.5   *Lattice energies of some metal dihalides at*
0 °K (*kcal/mole*)

Compound	Structure	Born cycle[a]	Equation (2.7)[b]
MgBr$_2$	CdI$_2$	572	511
MgI$_2$	CdI$_2$	553	474
CaI$_2$	CdI$_2$	494	453
MnF$_2$	Rutile	663	660
MnBr$_2$	CdI$_2$	583	520
MnI$_2$	CdI$_2$	569	480
PbF$_2$	Fluorite	596	582
PbI$_2$	CdI$_2$	515	458
ZnF$_2$	Rutile	717	686
CdF$_2$	Fluorite	663	632
CdI$_2$	CdI$_2$	582	475
HgF$_2$	Fluorite	660	624
HgCl$_2$	Layer	626	
HgBr$_2$	Layer	620	
HgI$_2$	Layer	623	

[a] See footnote (a) to table 2.2.   [b] See footnote (b) to table 2.2.

form to the definition possess properties (*a*), (*b*) and (*c*) and second, that, in a series of compounds, increasing differences between cycle and calculated lattice energies are frequently matched by departures from these properties. It has already been shown, as far as possible, that the first of these two demands is satisfied. The second will now be examined.

We shall first consider the lattice energies of the alkaline earth dihalides in tables 2.2 and 2.5. Except for magnesium chloride and bromide and the iodides of magnesium and calcium, these compounds all have the fluorite, rutile, slightly deformed rutile or

PbCl$_2$ lattices. All of these structures are three-dimensional and, in the first three, the coordination around the constituents is highly regular or nearly so. The alkaline earth dihalides with these arrangements have cycle lattice energies that differ from the extended calculation values by less than 10 kcal/mole in each case. The four exceptions possess the cadmium iodide or cadmium chloride structures in which the metal atoms are arranged in layers octahedrally coordinated to the halogens but each halogen and its three nearest metal neighbours form a tetrahedron. On the other side of the halogen, the nearest neighbours are three other halogens in an adjacent layer. This unsymmetrical arrangement would not be anticipated with a hard sphere ionic model. At the same time the cycle values for the lattice energies exceed those obtained from extended calculation methods by 40 kcal/mole or more, and again it seems that these differences are larger the heavier the halogen.

With the dihalides of the other metals, the discrepancies between the compressibility and cycle values are larger still. The fluorides all have the rutile or fluorite structures, but nevertheless, considerable differences are apparent for the zinc, cadmium and mercury compounds. Again, when the size of the anion is increased, layer structures are adopted and the discrepancies become greater. By both energetic and crystallographic standards, the properties of the mercuric halides are very unlike those expected of an ionic compound. Although the fluoride has the fluorite structure, the calculated and cycle lattice energies differ by over 40 kcal/mole. Furthermore, the cycle values for the three other halides lie within 6 kcal/mole or less than 1 per cent of each other. The chloride, bromide and iodide all form layer lattices and boil below 400°, while the specific conductance of mercuric chloride at the melting point is less than one thousandth that of any alkaline earth dihalide.

These facts show how compounds denied the description 'ionic' by the suggested definition display departures from properties (a), (b) and (c) which are usually taken as evidence of the presence of ions in crystals.

Another change which is discernible when cycle and calculated lattice energies in a series of compounds diverge, is loss of additivity of internuclear distance with an analogous series in which these lattice energies are more in agreement. When the two ways of obtaining lattice energies give substantially different results, it is nearly always found that the cycle value is the greater. Thus, increasing differences in a series of compounds should imply an increasing additional contribution to the binding that is absent in an analogous series where the cycle and calculated lattice energies agree. This should show itself in a shortening of the internuclear distance relative to the latter series and, if a comparison is possible, the additivity of this quantity between the two series should be destroyed.

With the silver halides, for example, the difference between the cycle and calculated lattice energies increases from 7 kcal/mole for the fluoride to 27 kcal/mole for the iodide and the values for the bromide and iodide are identical (see table 2.4). The fluoride, chloride and bromide have the rock salt structure and the inter-nuclear distances in NaCl and AgCl are nearly identical, but the value for silver fluoride is 0·15 Å longer than in the corresponding sodium compound while for silver bromide it is 0·11 Å less. This indicates that the internuclear separations in the silver halide lattices are not co-additive with those of the alkali metal halides because they are relatively shortened as the size of the anion is increased. Silver iodide has the tetrahedrally coordinated zinc blende structure. Using a hard sphere model, the cation becomes so small and the anion–anion repulsion in the rock salt lattice so large that, unlike sodium or even lithium iodide, four coordination is preferable.

Again, the values of the calculated and cycle lattice energies diverge more severely from the fluoride of manganese to the iodide than they do in the corresponding magnesium series; the cycle lattice energy of magnesium fluoride is over 30 kcal/mole greater than that of the corresponding manganese salt, but that of the iodide is nearly 20 kcal/mole less. At the same time, the interatomic distances in the fluorides are in the order Mn > Mg while in the iodides this sequence is reversed.

Differences in the rates of divergence of cycle and calculated lattice energies in certain series of compounds have some interest-ing chemical consequences. Thus as the size of the anion is increased, the lattice energies of the monohalides of copper, silver and gold diminish less rapidly than those of the alkali metals while those of the dihalides of cadmium, mercury and lead fall more slowly than those of an alkaline earth metal of comparable size.

This implies that, relative to either the alkali metal or alkaline earth series, there is an increase in the strength of the cation–anion binding as one descends the halogen group. It is presumably this increase that is reflected in the tendency of these metals to be Class B acceptors, that is to form complexes with stability constants in the order $I^- > Br^- > Cl^- > F^-$, rather than in the reverse sequence characteristic of the metals whose salts are better described by properties (a), (b) and (c). A more obvious example of this effect is the comparative excellence of silver and mercuric fluorides as fluorinating agents in halogen exchange reactions. This is discussed in §2.10.

Summarizing, the definition that was suggested earlier for an ionic compound seems to be consistent with the other properties expected for a solid which contains ions. Nevertheless, it has some unsatisfactory features. As noted earlier, a coincidence of the cycle and calculated lattice energies suggests nothing about the electron density distribution along the cation–anion axis. Indeed if this distribution were used as a criterion of an ionic compound, then, as experimental values of r_0 are used, and non-ionic contributions to the binding should both raise the lattice energy and shorten the internuclear distance, it is likely that equations such as (2.7) or (2.8) can to some extent accommodate these contributions (see Phillips and Williams 1965). Some support for this belief is provided by the inversion of the order of the lattice energies of magnesium and manganese halides between the fluoride and the iodide which was referred to above, for this inversion is also apparent in the values calculated from the Born–Landé equation. Both fluorides and both iodides are isostructural, while the repulsive exponents are also very similar. Consequently, the reversal of the order of internuclear distance in moving from the fluorides to the iodides causes a reversal in the order of lattice energies. It is very possible, therefore, that this reflects the self-compensating effect of the Born–Landé equation.

Another disadvantage of the definition is its limited applicability. The evaluation of cycle lattice energies is only possible when the heats of formation of the gaseous ions are known, and, in virtually all cases, this prevents classification of salts that might contain diatomic or polyatomic ions.

It would seem then that, until the analysis of electron density distributions becomes easier and more accurate, the word 'ionic' will remain more a label for a class of compounds with defined properties than an adjective bearing any precise suggestion of the net charges on the constituent atoms.

Currently, there is no entirely satisfactory rationalization of the discrepancies between compressibility and cycle lattice energies in a series of compounds. It will be noted from tables 2.2 and 2.5 that the differences between the cycle and calculated values for the iodides vary in the order Cd > Mn > Mg > Pb > Ca > Ba which is the order of the electron affinities (second ionization potential) of the gaseous dipositive ions. It is often suggested that the electron affinity, among other things, reflects the 'polarizing power' of the cation and that non-ionic contributions to the binding increase with this and the polarizability of the anion which can be obtained from refractive index measurements. While this observation is suggestive, there is no universally accepted way of measuring or even defining polarizing power, and the concept has, as yet, no quantitative applications.

2.5. Quantitative uses of the lattice energy.

The determination of electron affinities and related terms. In §2.2 and 2.3 it was shown that, for a number of alkali metal compounds, the compressibility equations discussed in §2.2 gave lattice energies which were little different from the Born–Haber cycle values. If it is assumed that this will be true of any alkali metal, and even of some alkaline earth metal salts, then the lattice energy of such a compound may often be calculated from a knowledge of its structure and compressibility behaviour. This figure can then be substituted in (2.2) and a value obtained for the enthalpy of formation of the gaseous anion. If the enthalpy of formation of the uncharged radical from its constituent elements is also available, then the electron affinity may be determined.

For example, extended calculation values for the lattice energies of some alkali metal and alkaline earth monoxides yield values of about 220 ± 20 kcal/mole for $\Delta H_f^0 [O^{2-} (g)]$. The dissociation energy of molecular oxygen is 119 kcal/mole, so the double electron affinity of oxygen atoms is about -160 kcal/mole. Since the absorption of the first electron has been experimentally shown to proceed with evolution of heat, it is the very unfavourable energy term associated with the uptake of the second that is the principal cause of the large positive value of $\Delta H_f^0[O^{2-}(g)]$. By similar calculations, the heats of formation of gaseous ions like $HF_2^-, CN^-, BH_4^-, BF_4^-$ and NO_3^- have been determined. These offer routes to the enthalpies of reactions like

$$BF_3(g) + F^-(g) \rightarrow BF_4^-(g) \qquad [2.3]$$

for

$$\Delta H^0 = \Delta H_f^0[BF_4^-(g)] - \Delta H_f^0[BF_3(g)] - \Delta H_f^0[F^-(g)]. \qquad (2.9)$$

All the quantities on the right-hand side are known, the last two by totally experimental measurements and the first by extended calculation of the lattice energy of KBF_4 followed by substitution in (2.2). The value of ΔH^0 obtained is -92 kcal/mole (Bills and Cotton, 1960).

If the cation rather than the anion is polyatomic, then its enthalpy of formation in the gas phase may also be calculated by this method. Ammonium chloride and bromide have the eight co-ordinate caesium chloride structure at room temperature, while the iodide possesses a rock salt lattice. Extended calculation methods yield the lattice energies shown in table 2.6 and, with the aid of (2.2), these give 141, 143 and 142 kcal/mole respectively for $\Delta H_f^0[NH_4^+(g)]$ if we assume that $U_{290}[NH_4X] = U_0[NH_4X]$. From the ionization potential and dissociation energy of hydrogen, $\Delta H_f^0[H^+(g)]$ is readily obtained and, with the heat of formation of ammonia, the figure -214 kcal/mole is derived for the standard enthalpy change of the reaction,

$$NH_3(g) + H^+(g) \rightarrow NH_4^+(g) \qquad [2.4]$$

at 298 °K. This will differ little from the internal energy change at 0 °K, so the proton affinity of ammonia is very large and positive.

The structural change in moving from ammonium iodide to

TABLE 2·6 *Lattice energies of ammonium halides at 0 °K (kcal/mole)*

	NH₄Cl	NH₄Br	NH₄I
	156	151	142

See Ladd and Lee (1964).

ammonium bromide implies that, on a hard sphere model, there is insufficient room for eight large iodide ions around the cation, but that this restriction is removed for the three other halogens. It is therefore surprising to find that ammonium fluoride has the rather open, tetrahedrally coordinated wurtzite structure. In addition, if the value of ΔH_f^0 [NH_4^+ (g)] derived above is used to calculate a cycle value for the lattice energy of the compound, the figure exceeds the one obtained by extended calculation methods by about 15 kcal/mole. Both these discrepancies can be understood if the existence of four tetrahedrally arranged, linear N—H————F hydrogen bonds emanating from each nitrogen is recognized. Their presence is apparent from the N—H rocking and stretching frequencies which are higher and lower respectively than in other ammonium salts, and they stabilize the wurtzite structure yielding larger lattice energies than expected. Ammonium fluoride is very unusual in having appreciable solubility in ice, whose hydrogen-bonded structure is closely related to it and with which it is isoelectronic.

The stabilities of hypothetical compounds. Because the lattice energies of metallic salts are so evidently dependent on internuclear distance, the value for an unknown compound of this type can, without undue error, be assumed equal to that of the analogous salt of a metal of similar atomic number. Its heat of formation may then be calculated from (2.2) and its instability with respect to likely decompositions assessed.

Thus putting the lattice energy of $TlCl_2$ equal to that of $PbCl_2$ at 534 kcal/mole,[1] we obtain ΔH_f^0 [$TlCl_2(s)$] = 5 kcal/mole. But experimentally it is found that ΔH_f^0 [$TlCl$] = $-48 \cdot 8$ kcal/mole, so for the reaction:

$$TlCl_2 \rightarrow TlCl + \tfrac{1}{2}Cl_2 \qquad\qquad\qquad [2.5]$$

$\Delta H^0 = -54$ kcal/mole. Using Latimer's method for estimating the entropy of solids, S^0[$TlCl_2$] = 33 cal/deg.mole and therefore $\Delta S^0 = +20$ cal/deg.mole. Consequently $\Delta G^0 = -60$ kcal/mole. This suggests that $TlCl_2$ is unstable to dissociation. Chlorination of TlCl does yield $TlCl_2$ but the diamagnetic product is probably analogous to $Ga^I Ga^{III} Cl_4$ which, from X-ray analysis, contains

---

[1] Ions like Tl^{2+} could conceivably pair spins and polymerize by forming metal–metal bonds (cf. Hg_2^{2+}). This would produce a limited additional stabilization that is neglected in such assumptions.

tetrahedral $GaCl_4^-$ ions. This observation is therefore compatible with the thermodynamic conclusion that the chloride of the dipositive state is unstable.

By assuming that their lattice energies equal those of the adjacent alkali metal compounds, the heats of formation of the halides and oxides of the noble gases, the alkaline earths and aluminium in the $+1$ oxidation state have been calculated (Waddington, 1959). The values suggest that the noble gas compounds are all highly unstable to the uncombined elements, while the remainder should disproportionate readily into the element and the halide or oxide of the metal containing closed shell ions.

Again, by using either the lattice energies of the adjacent alkaline earth compounds, the heats of formation of the dihalides and monoxides, MO, of the alkali metals may be estimated. All are highly unstable to the formation of the monohalides or oxides M_2O with loss of the appropriate halogen or oxygen.

If the problem of the instabilities of nearly all the hypothetical compounds mentioned so far in this section is restated by means of thermodynamic cycles involving lattice energies and ionization potentials, the analysis of the data shows that, if the estimated lattice energies are reasonable, the instability may be attributed to the large differences in ionization potential for the configurations $(n-1)p^6ns^2$, $(n-1)p^6ns^1$ and $(n-1)p^6$. For the ions of adjacent elements with the same charge, the ionization potential of the last configuration is always very much greater than that of the second which is considerably less than that of the first. Consequently the ionization potential variations dominate those in the stabilities of oxidation states, and ions or atoms with the electronic structure $(n-1)p^6$ tend to be very stable with respect to those with adjacent configurations. The ionization potentials are an indication of the stability of the octet which has long been an accepted part of inorganic chemistry.

As a further example of this use of the lattice energy, the reasons for the instability of ammonium hydride may be examined. From the cycle in fig. 2.4,

$$\Delta H^0 = U[\mathrm{NH_4H}] - x - D[\mathrm{H_2}] - I + E_a + 2RT \tag{2.10}$$

$$= U[\mathrm{NH_4H}] - x - \Delta H_f^0[\mathrm{H^+(g)}] - \Delta H_f^0[\mathrm{H^-(g)}] + 2RT, \tag{2.11}$$

3. The Alkali Metals and the Alkaline Earths

3.1. Introduction. In chapter 2, the lattice energies of alkali metal salts, obtained from (2.7) and (2.8), were shown to agree closely with the cycle values. For the salts of other metals, this agreement is less precise, but the problems discussed in §2.8–2.11 showed that the simplest Kapustinskii equation, which is derived from (2.7), could still be profitably applied to some aspects of their chemistry. These facts suggest that certain parts of the chemistry of the alkali metals and alkaline earths might be susceptible to a similar kind of treatment.

TABLE 3.1 *Some properties of the alkali metals*

	Li	Na	K	Rb	Cs
Ionic radius (Å)	0·68	1·00	1·33	1·47	1·68
L_v (kcal/mole)	38·5	25·9	21·5	19·6	18·7
I_1 (kcal/mole)	124	118	100	96	90
I_2 (kcal/mole)	1744	1091	734	634	579

TABLE 3.2 *Some properties of the alkaline earth metals*

	Be	Mg	Ca	Sr	Ba
Ionic radius (Å)	—	0·68	0·99	1·16	1·34
L_v (kcal/mole)	77·9	35·4	42·2	39·2	42·5
I_1 (kcal/mole)	215	176	141	131	120
I_2 (kcal/mole)	420	347	274	254	231
I_3 (kcal/mole)	3548	1848	1181	—	—

Some important general properties of these elements are shown in tables 3.1 and 3.2. The comparatively low first and very high second ionization potentials of the alkali metals are a quantitative demonstration of the stability of the electronic configuration ns^2np^6 with respect to the loss or gain of electrons. Calculations of where I and E_a are the ionization potential and electron affinity of hydrogen. As the internuclear distances in the rock salt lattices of RbI and NH_4I are very similar, the lattice energy of ammonium hydride at 25° may be put equal to that of rubidium hydride (161 kcal/mole). From the previous section, $x = -214$ kcal/mole while the enthalpies of formation of the gaseous proton and hydride ion are 367 and 33 kcal/mole respectively. Consequently $\Delta H^0 = -24$ kcal/mole. The molar entropies of gaseous ammonia and

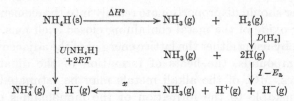

Fig. 2.4. Thermodynamic cycle for the decomposition for ammonium hydride.

hydrogen are 46·0 cal/deg.mole and 31·2 cal/deg.mole, while that of crystalline ammonium hydride may be estimated as about 16 cal/deg.mole by Latimer's method. Thus $\Delta S^0 = +61$ cal/deg.mole, $\Delta G^0 = -42$ kcal/mole and the compound is evidently very unstable to this dissociation. Compared with the ammonium halides, the salient feature in the cycle is the low electron affinity of hydrogen which is over 50 kcal/gm. atom less than that of iodine. The high dissociation energy of hydrogen is also of importance in a comparison with ammonium iodide. It is over 30 kcal/mole greater than that of HI.

The most sensational application of the method described in this section was the preparation by Bartlett of the compound $XePtF_6$. After isolating a compound PtO_2F_6 by the reaction of platinum hexafluoride with oxygen at room temperature, he discovered that it was isostructural with $KSbF_6$, and that the volumes of the two unit cells were very similar. This suggested that it should be formulated as $O_2^+PtF_6^-$ and that the spatial contribution of the cation to the lattice parameters differed little from that of K^+. Although the Xe^+ ion might have been expected to have a radius nearer Cs^+ than K^+, the difference in the lattice energies of $O_2^+PtF_6^-$ and $Xe^+PtF_6^-$ should in fact be small because of the very large anion. At the same time Bartlett observed that the first ionization potential of the oxygen molecule was almost identical with that of xenon. These facts suggested that $XePtF_6$ might be stable with respect to decomposition into xenon and platinum hexafluoride, and when the two gases were mixed at room temperature, the first noble gas compound was deposited as an orange-yellow solid on the walls of the containing vessel. Subsequent X-ray investigation has disclosed that the compound should probably not be formulated as $Xe^+PtF_6^-$. As this suggests that it should be more stable than expected, the value of Bartlett's calculation, from a predictive point of view, is unimpaired.

Thermodynamically, the stability of $XePtF_6$ may be attributed to the unknown, but undoubtedly very large electron affinity of platinum hexafluoride. Ions whose formation from their neutral species is energetically very unfavourable may also be stabilized by using the process [2.3], the large negative standard enthalpy of which was referred to earlier in this section. For the reaction,

$$XeBF_4(s) \rightarrow Xe(g) + BF_3(g) + \tfrac{1}{2}F_2(g), \qquad\qquad [2.6]$$

$$\Delta H_{298}^0 = U_{298}[XeBF_4] + 2RT + 92 - \Delta H_f^0[F^-(g)]$$
$$- \Delta H_f^0[Xe^+(g)] \quad (2.12)$$

$\Delta H_f^0[Xe^+(g)]$ is 281 kcal/mole, and by assuming that $U_{298}[XeBF_4]$ is 145 kcal/mole, about 7 kcal/mole less than the value calculated for KBF_4, we obtain $+22$ kcal/mole for ΔH_{298}^0. Unfortunately, the gaseous products would ensure that ΔS_{298}^0 was about 90 cal/deg.mole, so the standard free energy of decomposition would, from these calculations, be about -4 kcal/mole at 298 °K. However, by lowering the temperature, the influence of the entropy term could be reduced while any deviations from the ionic model should, if anything, raise the lattice energy and stabilize the compound. Its preparation is therefore a distinct possibility. The fluoroborates of several unusual polyatomic cations have been prepared. The direct reaction of F_2O_2, ClF_3 and ClO_2F with boron trifluoride yields the compounds $O_2^+BF_4^-$, $ClF_2^+BF_4^-$ and $ClO_2^+BF_4^-$, while $NO^+BF_4^-$ and $NO_2^+BF_4^-$ are also known. The anion SbF_6^- exerts a similar stabilizing influence on unusual cations, presumably because of the favourable nature of the process,

$$SbF_5(g) + F^-(g) \rightarrow SbF_6^-(g). \qquad\qquad [2.7]$$

The compounds $XeSbF_6$, $NF_4^+SbF_6^-$ and $ClO_2^+SbF_6^-$ have all been prepared. With all these products the main experimental criteria for the presence of ions are the high symmetry of the discrete anion and a change in the geometry of the cation relative to the neutral radical. These last two properties are usually detected by vibrational spectroscopy or X-ray analysis.

2.6. Ionic radii. In § 2.1 it was pointed out that a number of compounds with the characteristic ionic properties have inter-nuclear distances which are approximately additive. This suggests the possibility of assigning radii to individual ions. Apart from their ability to reproduce, with reasonable accuracy, a large number of internuclear distances, such radii give a very useful idea of the variations in the lattice energies of crystals. In addition, if they are assigned on a hard sphere model, they can be used to interpret, with some success, the variations in structure which occur in crystals as the ratio of the cation and anion radii alters.

Because the internuclear distances in ionic compounds vary somewhat with the coordination number, the standard compila-tions of ionic radii usually contain figures for a value of six. Those in table 2.7 have been compiled in the following way: It is evident from a study of internuclear distances that the contribution of the cation diminishes very rapidly as its charge increases, while that of the anion, if anything, increases slightly. This suggests that cations are in general smaller than anions, cations being smaller than their parent atoms while anions are a little bigger. If a crystal is selected in which the anion is likely to be very much larger than the cation, then using a hard sphere model, it is possible that the internuclear distance will be determined by anion–anion rather than cation–cation contacts.

The internuclear distance in the rock salt structure of lithium chloride is $2 \cdot 570$ Å. By assuming that the anions are just in con-tact, a value of $2 \cdot 570/\sqrt{2}$ or $1 \cdot 82$ Å was obtained for the radius of the chloride ion. The comparatively large difference between the inter-nuclear distances in the fluorides and chlorides of lithium in table 2.1 is at least consistent with this assumption, and the value of the anion radius agrees well with that obtained by Waddington (1966) by a variety of methods. Using it, the radii of the ions in the fluorides, chlorides and bromides of sodium, potassium and

TABLE 2.7   *Ionic radii*

Ion	Radius (Å)	Ion	Radius (Å)	Ion	Radius (Å)
Li^+	0·68	Co^{2+}	0·71	Sm^{3+}	0·93
Na^+	1·00	Ni^{2+}	0·67	Eu^{3+}	0·91
K^+	1·33	Pd^{2+}	0·83	Gd^{3+}	0·90
NH_4^+	1·44	Al^{3+}	0·50	Tb^{3+}	0·88
Rb^+	1·47	In^{3+}	0·76	Dy^{3+}	0·87
Cs^+	1·68	Tl^{3+}	0·84	Ho^{3+}	0·86
Mg^{2+}	0·68	Ti^{3+}	0·63	Er^{3+}	0·84
Ca^{2+}	0·99	V^{3+}	0·59	Tm^{3+}	0·83
Sr^{2+}	1·16	Cr^{3+}	0·58	Yb^{3+}	0·82
Ba^{2+}	1·34	Mn^{3+}	0·60	Lu^{3+}	0·81
Ra^{2+}	1·43	Fe^{3+}	0·61	H^-	1·48
Ag^+	1·13	Co^{3+}	0·58	F^-	1·33
Zn^{2+}	0·70	Sc^{3+}	0·69	Cl^-	1·82
Cd^{2+}	0·97	Y^{3+}	0·86	Br^-	1·98
Hg^{2+}	1·03	La^{3+}	1·02	I^-	2·20
Pb^{2+}	1·21	Ce^{3+}	1·00	O^{2-}	1·42
Eu^{2+}	1·16	Pr^{3+}	0·97	S^{2-}	1·84
Mn^{2+}	0·80	Nd^{3+}	0·95	Se^{2-}	1·97
Fe^{2+}	0·75	Pm^{3+}	(0·94)	Te^{2-}	2·17

rubidium were calculated from the internuclear distances, the departures from additivity being minimized by the method of Waddington (1966). The radii of the ions Li^+, NH_4^+, Cs^+, Ag^+ and H^- were then obtained from the internuclear distances in the rock salt structure of LiF, NH_4I, CsF, AgF and the alkali metal hydrides.

In halide crystals containing multiply charged cations there is no question of equal coordination throughout, and radii for these cations have been obtained as in the past by selecting compounds where the cation is six coordinate and ignoring the effects of the different coordination of the anions on the internuclear distance. Furthermore, because the additivity relationship between the interatomic distances in a series of salts of a particular cation and those in the corresponding series of an alkali metal or an alkaline earth begins to disappear as the anion becomes larger and the compressibility equations begin to give low values for the lattice energy, it is advisable to determine radii from fluorides and oxides. The value for the oxide ion in table 2.7 was determined by taking the mean difference in the cation–anion separations in the oxides and fluorides of Mg^{2+}, Mn^{2+}, Co^{2+} and Ni^{2+}. These compounds all have six coordination around the cation, and the octahedron is only very slightly distorted in the rutile structure of the fluorides. An identical value is obtained by assuming anion–anion contact in MgS and calculating the mean difference between the radii of the oxide and sulphide anions from the distances in the oxides and sulphides of calcium, strontium and barium. These seven compounds have the rock salt structure. Save for those of Hg^{2+}, Pb^{2+}, Ra^{2+}, Eu^{2+} and Pd^{2+}, the remaining cation radii in table 2.7 were then obtained from the mean interatomic separations in six co-ordinate oxides or fluorides, although in some cases, notably the cubic rare earth oxide structure, this coordination is far from octahedral. The exceptions were calculated from the differences between the interatomic distances in their difluorides and in the difluorides of calcium, strontium and barium. These compounds all have the fluorite structure. Values for S^{2-}, Se^{2-} and Te^{2-} were obtained from the calcium, strontium and barium compounds with the rock salt lattice. The values in table 2.7 are similar to those given by Pauling (1960).

By using these figures which were derived using a hard sphere model, it is possible to understand why, for example, difluorides adopt the rutile rather than the fluorite structure when the radius of a dipositive ion is less than about 0·95 Å. Regarding the ions as hard spheres, it is energetically profitable to maximize the number of cation–anion contacts, but eight coordination of any ion is only possible if the ratio of its radius to the radius of the other is greater than $(\sqrt{3}-1)$ (see Evans, 1964). Such an approach, however, has only a limited success. It predicts for example that potassium and rubidium fluorides should have the eight coordinate caesium chloride structure, while in lithium bromide and lithium iodide the cations should be tetrahedrally coordinated as in the zinc blende or wurtzite lattices.

The energy factors which determine the taking up of a particular coordination are evidently more subtle than a hard sphere model implies. In particular, the increase in the Madelung constant[1]

[1] This increase is in any case, small. Compare, for example, the values for the wurtzite, sodium chloride and caesium chloride structures in table 2.8 (p. 39).

caused by a rise in the number of cation–anion contacts is, in energetic terms, offset by an increase in the internuclear distance. This has a fortunate result which will be discussed in the next section.

In conclusion, little absolute significance should be attached to the ionic radii in table 2.7. This has been emphasised by the very careful X-ray diffraction measurements which show that, in the limited cases studied, the minimum in the electron density distribution may be as much as 0·3 Å further from the cation than the ionic radius implies. The principal use of these radii lies merely in their capacity to roughly reproduce internuclear distances and those properties to which the latter is related.

2.7. The Kapustinskii equations. It is unfortunate that the lattice energy equations discussed in §2.2 are dependent upon the Madelung constant which is a characteristic of the crystal structure. For comparative purposes, it would be very useful to have some fairly precise understanding of the way in which the lattice energy varies with the size of the constituent ions, regardless of structural alterations. Kapustinskii (1933 and 1956) pointed out that, if the Madelung constants for a number of structures were divided by ν, the number of ions in one molecule, the values obtained were almost constant, varying between 0·74 for the cuprite and 0·88 for the caesium chloride structure (table 2.8). In addition, the order of values was roughly that of the average coordination number around the ions, and since the equilibrium internuclear distance in any ionic compound generally increases

TABLE 2.8   *Variation of M/ν with coordination number (see text)*

Structure	M	M/ν	Average coordination number
Caesium chloride (M^+X^-)	1·763	0·88	8
Sodium chloride (M^+X^-)	1·748	0·87	6
Fluorite ($M^{2+}X_2^-$)	2·519	0·84	$5\frac{1}{3}$
Wurtzite (M^+X^-)	1·641	0·82	4
Rutile ($M^{2+}X_2^-$)	2·408	0·80	4
Anatase ($M^{2+}X_2^-$)	2·400	0·80	4
Cuprite ($M_2^+X^{2-}$)	2·221	0·74	$2\frac{2}{3}$

slightly with coordination number, the proportional variations in $M/r_0\nu$ will be even less than those in M/ν. Consequently the value of the former is almost constant whatever the structure, and the value for any one will serve for all. For the six-coordinate sodium chloride lattice, $M/r_0\nu$ is $0\cdot874/r_0$ and we may replace r_0 by $(r_+ + r_-)$, where r_+ and r_- are the cation and anion radii for six coordination in table 2.7. Making this substitution in (2.7), we have,

$$U_0 = \frac{0\cdot874 N_0 \nu z_+ z_- e^2}{r_+ + r_-}\left(1 - \frac{1}{n}\right). \tag{2.13}$$

As the repulsive term is only eight to twenty per cent of the coulombic, Kapustinskii proposed the substitution of an average value of $n = 9$. Converting to kcal/mole:

$$U_0 = \frac{256\nu z_+ z_-}{r_+ + r_-}. \tag{2.14}$$

For the rock salt structures of the alkali metal halides, ρ in (2.8) is almost constant at $0\cdot345$. This, with the substitution for M/r_0 gives in kcal/mole:

$$U_0 = \frac{287\cdot2\nu z_+ z_-}{r_+ + r_-}\left(1 - \frac{0\cdot345}{r_+ + r_-}\right) \tag{2.15}$$

(2.15) is more accurate than (2.14) because possible variations in the second term in the brackets are partially covered by including the crystal radii [n in (2.7) generally increases with the size of the ions]. It has been used extensively by the Russian School for the calculation of lattice energies in the absence of structural and compressibility data. A further development has been the assignment of 'thermochemical radii'. If, for example, a complex anion X^- forms compounds M_1X and M_2X with two different cations M_1^+ and M_2^+, then by subtraction of the equations (2.2) for both compounds:

$$U[M_1X] - U[M_2X] = \Delta H_f^0[M_1^+ (g)] - \Delta H_f^0[M_2^+ (g)]$$
$$- (\Delta H_f^0[M_1X] - \Delta H_f^0[M_2X]). \tag{2.16}$$

If the enthalpies of formation of the compounds and the gaseous cations are known, the difference in the lattice energies can be found. By equating this to the difference in the Kapustinskii expressions (2.15), the radius of X^- may be calculated by using those of M_1^+ and M_2^+. Some values obtained in this way by Russian workers are shown in table 2.9. Their variations are roughly what one would expect from a knowledge of those in bond lengths and the radii of monatomic ions, but it must be emphasised that they have been obtained in a different way from those in table 2.7. They may of course be used to obtain approximate values of lattice energies if the radius of the cation in the compound is available for substitution in (2.15).

TABLE 2.9 *Thermochemical radii of some anions*

Ion	Radius (Å)	Ion	Radius (Å)	Ion	Radius (Å)
NH_2^-	1·30	NO_3^-	1·89	O_2^{2-}	1·80
OH^-	1·40	BrO_3^-	1·91	CO_3^{2-}	1·85
NO_2^-	1·55	ClO_3^-	2·00	SO_4^{2-}	2·30
CH_3COO^-	1·59	ClO_4^-	2·36	CrO_4^{2-}	2·40
HCO_3^-	1·63	MnO_4^-	2·40	MoO_4^{2-}	2·54
IO_3^-	1·82	IO_4^-	2·49	PO_4^{3-}	2·38
CN^-	1·82	BF_4^-	2·28	AsO_4^{3-}	2·48

The incorporation of a constant value for the repulsive term makes (2.14) highly unsuitable for the calculation of absolute values of lattice energies. Again, if there is anion–anion contact, replacement, of r_0 by $(r_+ + r_-)$ is another serious approximation. Nevertheless, (2.14) does give a reasonable idea of the way in which the coulombic term varies with ionic size. On an ionic model, it is almost invariably the variations in the coulombic terms that determine those in a series of lattice energies. When this situation arises, the simplicity of (2.14) makes it highly suitable for comparative purposes, and the conclusions obtained by its use often hold even when the compounds involved are far from ionic. In the remainder of this book, the small correction from 0–298 °K is ignored, and (2.14) is used extensively as an expression for the lattice energy at normal temperatures.

2.8. The stabilization of high oxidation states by fluorine and oxygen.
Halides containing the higher oxidation states of metals are usually fluorides rather than chlorides, bromides

or iodides. Such compounds usually decompose to the halide of a lower oxidation state as opposed to their constituent elements so, when discussing their stabilities, it is more important to consider the first of these two dissociations. Consider a solid salt MX_{n+1} decomposing into another, MX_n, and $\frac{1}{2}X_2$. From fig. 2.5[1] and (2.14)

$$\Delta H^0 = U[MX_{n+1}] - U[MX_n] - I_{n+1} - \Delta H_f^0[X^-(g)] - \tfrac{3}{2}RT \quad (2.17)$$

$$= \left[\frac{256(n+1)(n+2)}{r(M^{n+1}) + r(X^-)} - \frac{256n(n+1)}{r(M^{n+}) + r(X^-)}\right] - I_{n+1}$$
$$- \Delta H_f^0[X^-(g)] - 0 \cdot 9. \quad (2.18)$$

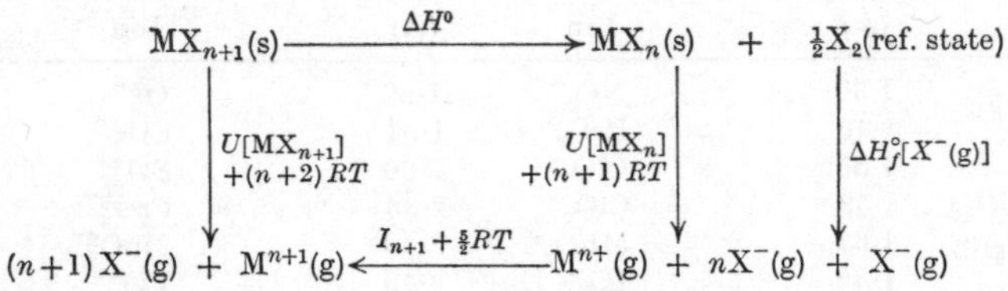

Fig. 2.5. Thermodynamic cycle for the decomposition of a halide MX_{n+1}.

The data in table 2.10 shows that the term $-\Delta H_f^0[X^-(g)]$ favours the stabilization of the halides in the order $F > Cl > Br > I$, the pre-eminent position of fluorine being due to the low dissociation energy of the molecule. In addition, if the term in the bracket is put equal to x while n and the cation radii are kept constant,

$$\frac{dx}{dr(X^-)} = -256(n+1)\left[\frac{n+2}{[r(M^{n+1}) + r(X^-)]^2} - \frac{n}{[r(M^{n+}) + r(X^-)]^2}\right].$$
$$(2.19)$$

Although it is not necessarily crucial, $r(M^{n+1})$ will be less than $r(M^{n+})$. Consequently, as n is positive $\dfrac{dx}{dr(X^-)}$ will be less than zero and x will become more negative as the size of the anion is increased. Now the changes in $T\Delta S^0$ from halogen to halogen are small compared with those in ΔH^0. Thus the variations in both $-\Delta H_f^0[X^-(g)]$ and in the difference in the lattice energies place the capacities to

---

[1] Any gaseous electrons have been omitted from the equations, so that some steps in the cycle are not balanced with respect to charge.

TABLE 2.10 *Enthalpies of formation of gaseous halide anions at* 298 °K (*kcal/mole*)

	F	Cl	Br	I
$\frac{1}{2}\Delta H_f^0 [X_2 (g)]$	0	0	3·7	7·46
$\frac{1}{2}D(X_2)$	18·9	29·1	23·0	18·08
$-E_a[X]$	−82·1	−86·5	−81·1	−71·0
$-\frac{5}{2}RT$	−1·5	−1·5	−1·5	−1·5
$\Delta H_f^0 [X^-(g)]$	−64·7	−58·9	−55·9	−47·0

stabilize the higher oxidation state in the order $F > Cl > Br > I$. For example, the fluorides are the only halides of dipositive silver, tripositive cobalt and manganese and tetrapositive manganese, cerium, praseodymium and terbium to have been prepared. Again, the iodides of copper(II) and iron(III) have not been obtained pure because they are unstable at room temperature to CuI and FeI_2. In contrast, all the remaining halides of these two oxidation states are known.

The only other element that rivals fluorine in its ability to stabilize high oxidation states is oxygen. The importance of investigating the correct decomposition reaction when discussing the instability of a substance may be illustrated by comparing the stabilities of the compounds MF_n and $MO_{n/2}$ with respect to their elements. With a suitable adjustment of (2.2), we have,

$$\Delta H_f^0[MF_n] = \Delta H_f^0[M^{n+}(g)] + n\Delta H_f^0[F^-(g)] - U[MF_n]$$
$$- (n+1)RT, \quad (2.20)$$

$$\Delta H_f^0[MO_{n/2}] = \Delta H_f^0[M^{n+}(g)] + n/2\,\Delta H_f^0[O^{2-}(g)]$$
$$- U[MO_{n/2}] - (n/2+1)RT. \quad (2.21)$$

and ignoring the small RT terms,

$$\Delta H_f^0[MF_n] - \Delta H_f^0[MO_{n/2}] = n\Delta H_f^0[F^-(g)]$$
$$- n/2\,\Delta H_f^0[O^{2-}(g)] - U[MF_n] + U[MO_{n/2}]. \quad (2.22)$$

The values of $\Delta H_f^0[F^-(g)]$ and $\Delta H_f^0[O^{2-}(g)]$ are -65 and 220 kcal/mole respectively. The large difference is mainly due to the higher dissociation energy of the oxygen molecule and the fact that the uptake of a second electron by the O^- ion is a very unfavourable process. Thus,

$$\Delta H_f^0[MF_n] - \Delta H_f^0[MO_{n/2}]$$
$$= -175n - U[MF_n] + U[MO_{n/2}]. \quad (2.23)$$

Using (2.14), neglecting the changes in the lattice energies between $0\ ^\circ K$ and $298\ ^\circ K$, and substituting $1\cdot4$ Å for the radii of the anions which are similar in size,

$$\Delta H_f^0[MF_n] - \Delta H_f^0[MO_{n/2}] = -175n + \frac{256n}{r(M^{n+}) + 1\cdot4}. \quad (2.24)$$

According to this relation, $\Delta H_f^0[MO_{n/2}]$ will be less than $\Delta H_f^0[MF_n]$ only if $r(M^{n+})$ is less than about $0\cdot1$ Å. If the simple ionic model is valid, this impossibly small value of $r(M^{n+})$ suggests that $-\Delta H_f^0[MF_n]$ should always exceed $-\Delta H_f^0[MO_{n/2}]$ by a substantial amount. That this conclusion is correct can be seen from the $+\Delta H_f^0$ values in table 2.11.

TABLE 2.11   *Standard enthalpies of formation of oxides and fluorides in kcal/mole*

AgF	$-48\cdot5$	LiF	$-145\cdot7$	HgF_2	-100	AlF_3	$-359\cdot5$	PbF_4	$-225\cdot1$
$\frac{1}{2}Ag_2O$	$-3\cdot6$	$\frac{1}{2}Li_2O$	$-71\cdot2$	HgO	$-21\cdot7$	$\frac{1}{2}Al_2O_3$	$-200\cdot3$	PbO_2	$-66\cdot3$

In a comparison of the stabilities of the two substances with respect to their elements, the relevant enthalpy terms are really $-\Delta H_f^0[MF_n]$ and $-\Delta H_f^0[M_2O_n]$ for these are closely related to values of $-\Delta G_f^0$ which correspond to equilibrium constants in which the pressures of oxygen and fluorine are raised to the same power.[1] However, it is when the values of $-\Delta H_f^0$ are small that they are chemically important, because only then will the compound concerned decompose at reasonably accessible temperatures. As $-\Delta H_f^0[MO_{n/2}]$ is always substantially less than $-\Delta H_f^0[MF_n]$, $-\Delta H_f^0[M_2O_n]$ is also less than $-\Delta H_f^0[MF_n]$ when $-\Delta H_f^0[MO_{n/2}]$ is small. This too is clear from the figures in table 2.11.

Low values of $-\Delta H_f^0$ are encountered among the halides and

[1] The entropies of dissociation will then be similar, and the $-\Delta G_f^0$ or $-\Delta H_f^0$ values will run in the same order as the decomposition temperatures (see §3.1).

oxides of the noble metals where these conclusions are chemically relevant. Thus platinum tetrafluoride is prepared by the fluorination of red-hot platinum, but the oxide $PtO_2 . H_2O$ cannot be dehydrated at 200–400° without decomposition to the metal. At 200°, silver oxide loses oxygen, but the fluoride melts unchanged at 435°. Mercuric fluoride boils at 650° but the oxide is reduced to the element by heating above 400°. These facts support the conclusion that, when kinetic factors are ignored, an ionic oxide that decomposes to its elements within commonly attainable temperature ranges up to about 1000°, should do so before the corresponding fluoride.

In fact, the stabilization of fluoride or oxides containing high oxidation states is governed by factors more subtle than those implied by (2.20) or (2.21). The instability of such compounds is usually registered by a decomposition into the oxide or fluoride of a lower oxidation state rather than into the elements. Under these circumstances, it is the difference in the lattice energies of the reactants and products that are important, not any one absolute value, and there is little to choose between the stabilizing effects of the two anions. Thus it is that high oxidation states, such as Ag^{II}, Co^{III}, Pb^{IV} and Ce^{IV}, are found as both oxides and fluorides.

2.9. The stabilization of low oxidation states by large anions. The instability of the lower or medium oxidation states of metals is often associated with disproportionation reactions of which the change,

$$3MX_2(s) \rightarrow M(s) + 2MX_3(s), \qquad\qquad [2.8]$$

is typical. The variations in the feasibility of this process as the size of the anion changes may be examined by means of the cycle in fig. 2.6. Here small terms in RT have been omitted, and

$$\Delta H^0 = 3U[MX_2] - 2U[MX_3] + 2I_3 - I_1 - I_2 - L_v, \qquad (2.25)$$

where L_v is the latent heat of sublimation of the metal. Using (2.14),

$$\Delta H^0 = 256 \left[\frac{18}{r(M^{2+}) + r(X^-)} - \frac{24}{r(M^{3+}) + r(X^-)} \right] + 2I_3$$
$$- I_1 - I_2 - L_v, \quad (2.26)$$

$$\frac{d(\Delta H^0)}{dr(X^-)} = -256 \left[\frac{18}{[r(M^{2+}) + r(X^-)]^2} - \frac{24}{[r(M^{3+}) + r(X^-)]^2} \right]. \quad (2.27)$$

Assuming that $r(M^{3+}) < r(M^{2+})$, then $d\Delta H^0/dr(X^-)$ is clearly positive. As the entropy change remains almost constant when the anion is changed, the compound of the lower oxidation state becomes more stable as the size of X^- is increased.

Thus TiF_2, which is unknown at room temperature, would probably decompose according to the equation

$$3TiF_2 \rightarrow Ti + 2TiF_3. \tag{2.9}$$

Similarly PtF_2 would probably yield the metal and PtF_4.

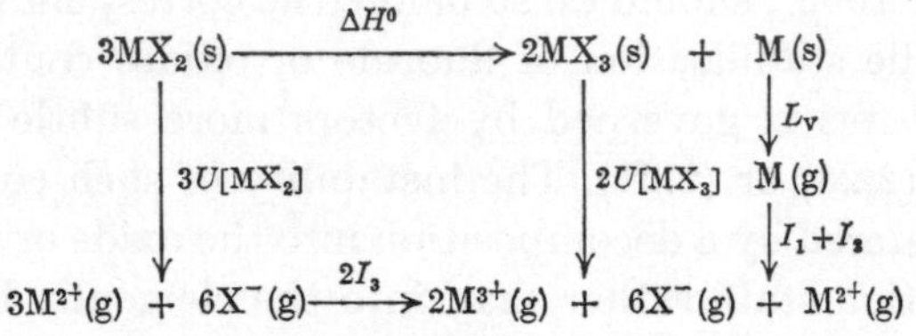

Fig. 2.6. Thermodynamic cycle for the disproportionation of a halide MX_2 (see footnote page 42).

The non-existence of these compounds contrasts with the stability of the three other halides of both elements in these oxidation states.

In the general case, the reaction becomes,

$$hMX_l(s) \rightarrow (h-l)M(s) + lMX_h(s), \tag{2.10}$$

where $h > l > 0$. Here it can be shown that,

$$\Delta H^0 = hU[MX_l] - lU[MX_h]$$

$$+ l\sum_{l}^{h} I_n - (h-l)\sum_{0}^{l} I_n - (h-l)L_v \tag{2.28}$$

and

$$\frac{d(\Delta H^0)}{dr(X^-)} = -256\left[\frac{hl(l+1)}{[r(M^{l+})+r(X^-)]^2} - \frac{hl(h+1)}{[r(M^{h+})+r(X^-)]^2}\right]. \tag{2.29}$$

Again, this simplified ionic model suggests that it is the large anions which stabilize intermediate oxidation states to disproportionation. For example, at room temperature the fluorides are the only unknown halides of gold(I) and copper(I), presumably because they are unstable to the formation of Au and AuF_3, and Cu and CuF_2 respectively.

The halides of gallium(I) are difficult to obtain pure because of their tendency to disproportionate, while those of the ion Cd_2^{2+} are unknown in the solid state. However, the reduction of gallium trichloride with the correct quantity of the metal yields

$$Ga^I(Ga^{III}Cl_4),$$

while from a solution of cadmium and $AlCl_3$ in molten cadmium dichloride, the compound $Cd_2^{2+}(AlCl_4^-)_2$ may be isolated. It is presumably the large anions that stabilize these unusual oxidation states to disproportionation.

The size dependence in a halide series may be perturbed if the higher halides show a severe increase in the differences between cycle and calculated lattice energies as the anion size is raised. $U_0[HgI_2]$, for example, exceeds $U_0[HgBr_2]$. This unexpected stabilization of the di-iodide makes Hg_2I_2 less stable to disproportionation than either Hg_2Br_2 or Hg_2Cl_2.

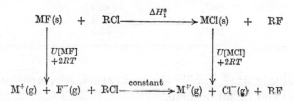

Fig. 2.7. Thermodynamic cycle for the fluorinating action of a metallic fluoride.

2.10. Halogen exchange reactions. The substitution of fluorine for another halogen is a particularly important process in the preparation of fluorine compounds. The problem of selecting the best fluorinating agent may be investigated with an ionic model by using the cycle in fig. 2.7. For an ionic salt, MF, reacting with a chloride RCl:

$$MF(s) + RCl \rightarrow MCl(s) + RF, \qquad [2.11]$$

the variations in the standard enthalpy, ΔH_1^0, as the cation, M^+, changes are given by

$$\Delta H_1^0 = U[MF] - U[MCl] + \text{constant}. \qquad (2.30)$$

The difference between the two lattice energies falls as the size of the cation increases, so the exchange reaction is more favourable when the cation is large. This agrees with the data in table 2.12 where some values of $\frac{1}{n}(\Delta H_f^0[MCl_n] - \Delta H_f^0[MF_n])$ are recorded. Caesium fluoride is a more powerful fluorinating agent than the corresponding potassium, sodium or lithium compounds.

TABLE 2.12   *The thermodynamics of the substitution of chlorine by fluorine*

Fluoride	$\frac{1}{n}(\Delta H_f^0[MCl_n] - \Delta H_f^0[MF_n])$ (kcal/mole)
AgF	18·1
HgF_2	23
CsF	23·4
KF	30·3
NaF	38·1
BaF_2	40·7
SbF_3	42·5
LiF	49·7
MgF_2	55·9

If a similarly sized, doubly charged cation, A^{2+}, is substituted for M^+, [2.11] and (2.30) become,

$$\tfrac{1}{2}AF_2(s) + RCl \rightarrow \tfrac{1}{2}ACl_2(s) + RF \qquad [2.12]$$

and

$$\Delta H_2^0 = \tfrac{1}{2}U[AF_2] - \tfrac{1}{2}U[ACl_2] + \text{constant}. \qquad (2.31)$$

As the constants in (2.30) and (2.31) are identical,

$$\therefore \Delta H_2^0 - \Delta H_1^0 = 256\left[\frac{3}{r(A^{2+}) + r(F^-)} - \frac{2}{r(M^+) + r(F^-)}\right.$$
$$\left. - \frac{3}{r(A^{2+}) + r(Cl^-)} + \frac{2}{r(M^+) + r(Cl^-)}\right]. \qquad (2.32)$$

Substituting $r(M^+)$ for $r(A^{2+})$,

$$\Delta H_2^0 - \Delta H_1^0 = 256\left[\frac{1}{r(M^+) + r(F^-)} - \frac{1}{r(M^+) + r(Cl^-)}\right] \qquad (2.33)$$

which is clearly positive. Thus, if the cations in two ionic fluorides have similar sizes, the compound containing the one of lower charge should be the more powerful fluorinating agent. That this conclusion is in accord with the facts may be seen by comparing the positions of lithium and magnesium fluorides, and potassium and barium fluorides in table 2.12. In practice, potassium fluoride is often preferred to the rubidium and caesium compounds because it is cheaper and picks up water less readily, e.g.

$$CH_2ClCH_2OH + KF \rightarrow CH_2FCH_2OH + KCl, \qquad [2.13]$$

$$C_2H_5SO_2Cl + KF \rightarrow C_2H_5SO_2F + KCl, \qquad [2.14]$$

$$[2.15]$$

Thermodynamically speaking, the most efficient fluorinating agents are the fluorides of these metals whose halides show a sharply increasing departure from the properties expected of an ionic compound as one descends the halogen group. Relative to an analogous ionic series, this departure creates an additional stabilization that increases from the fluoride, through the chloride and bromide, to the iodide and makes the exchange reaction more favourable. The difference between the cycle and calculated lattice energies increases by about 10 kcal/mole in passing from silver fluoride to silver chloride. Thus silver fluoride is a much more powerful fluorinating agent than sodium fluoride, a compound containing a singly charged cation of similar size whose halide lattice energies agree with calculated values. Silver fluoride does in fact head the list in table 2.12.

A particularly severe increase in departures from the ionic model was noted in the mercuric halide series. This enables HgF_2 to rival AgF in its thermodynamic fluorinating ability, despite the presence of a dipositive ion in the lattice. Both compounds are used for the replacement of chlorine and other halogens by fluorine, although in recent years AgF has been superceded because of the rather inert mixed halides that are formed at an intermediate stage. Some examples of the fluorinating action of HgF_2 which remains a very popular reagent, and of AgF are as follows:

$$SeCl_4 + 4AgF \rightarrow SeF_4 + 4AgCl, \qquad\qquad\qquad [2.16]$$

$$C_8H_{17}I + AgF \rightarrow C_8H_{17}F + AgI, \qquad\qquad\qquad [2.17]$$

$$S(CH_2CH_2Cl)_2 + 2AgF \rightarrow S(CH_2CH_2F)_2 + 2AgCl, \qquad [2.18]$$

$$CHBr_3 + \tfrac{3}{2}HgF_2 \rightarrow CHF_3 + \tfrac{3}{2}HgBr_2, \qquad\qquad\qquad [2.19]$$

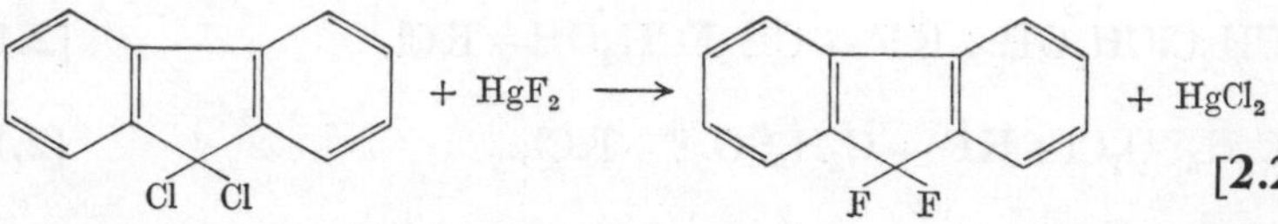

$$[2.20]$$

2.11. The stability of halides containing protonated bases.

The elements of Group V form compounds which contain the ions AH_4^+, and (2.14) throws some light on the factors that govern the relative stabilities of the halides. Halides of this type are unstable to the reaction:

$$AH_4X(s) \rightarrow AH_3(g) + HX(g), \qquad\qquad\qquad [2.21]$$

and in the cycle in fig. 2.8, x is an enthalpy term referring to the process shown, and y is the standard enthalpy of the reaction:

$$HX(g) \rightarrow H^+(g) + X^-(g). \qquad\qquad\qquad [2.22]$$

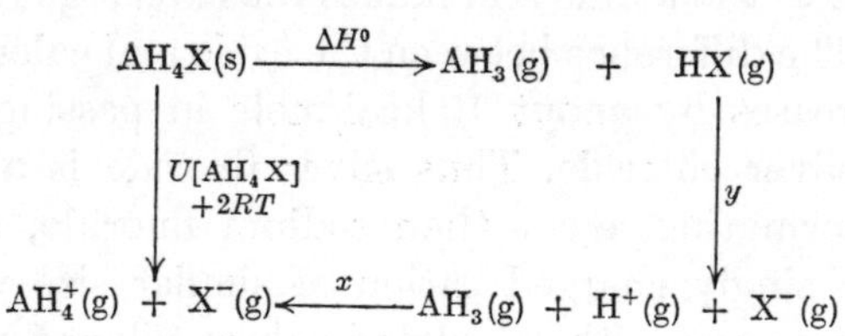

Fig. 2.8. Thermodynamic cycle for the decomposition of the halides of protonated hydrides of Group V.

From fig. 2.8,

$$\Delta H^0 = U[AH_4X] - x - y + 2RT \qquad\qquad (2.34)$$

and in comparing halides of the same cation, the equation reduces to

$$\Delta H^0 = U[AH_4X] - y + \text{constant}. \qquad\qquad (2.35)$$

The quantity y is shown in table 2.13 for each hydrogen halide, and we shall first compare the relative stabilities of the fluorides and iodides. The differences in the value of y for HF and HI is $53 \cdot 4$ kcal/mole, so that ΔH^0 is greater for the iodide if,

$$U[AH_4F] - U[AH_4I] < 53 \cdot 4, \tag{2.36}$$

i.e. if

$$512 \left[\frac{1}{r(AH_4^+) + 1 \cdot 33} - \frac{1}{r(AH_4^+) + 2 \cdot 20} \right] < 53 \cdot 4. \tag{2.37}$$

TABLE 2.13 ΔH_{298}^0 *values for the reaction* $HX\,(g) \to H^+\,(g) + X^-\,(g)$ *in kcal/mole*

HF	HCl	HBr	HI
$367 \cdot 3$	$330 \cdot 4$	$320 \cdot 0$	$313 \cdot 9$

The solution of the simple inequality shows that this will be true if $r(AH_4^+)$ is greater than about $1 \cdot 1$ Å. Because the sizes of the cations would be expected to increase down the group and the crystal radius of NH_4^+ is about $1 \cdot 44$ Å, this implies that, provided the entropy of dissociation remains roughly constant, the iodide should be the more stable of the two salts for all the cations, AH_4^+, of Group V. Similar calculations suggest that the chloride is more stable than the fluoride if $r(AH_4^+) > 1 \cdot 0$ Å and that the iodide is more stable than the bromide if $r(AH_4^+) > 2 \cdot 2$ Å. Unfortunately this last inequality is unreliable because the difference in y for HBr and HI is small and has an uncertainty of about ± 4 kcal/ mole caused by the uncertainties in the electron affinities of the two anions. Thus if the difference were 8 kcal/mole instead of $6 \cdot 1$ kcal/mole, the condition for higher stability of the iodide would be $r(AH_4^+) > 1 \cdot 6$ Å.

The date in table 2.14 show that these predictions are almost entirely fulfilled. For the ammonium halides the entropy term is sufficiently constant for the variations in ΔH^0 to reflect those in ΔG^0, and the order of stabilities is I < Br > Cl > F. With the phosphonium halides the non-existence of PH_4F may be reconciled

TABLE 2.14 *Standard enthalpies and free energies of decomposition of phosphonium and ammonium halides in kcal/mole*

	NH_4F	NH_4Cl	NH_4Br	NH_4I
ΔH^0	35·1	42·1	45·0	43·5
ΔG^0	14·1	21·8	25·2	23·5
	PH_4Cl	PH_4Br	PH_4I	
ΔH^0	14·0	23·1	24·3	
ΔG^0	− 7·5[a]	1·8	3·5	

[a] Estimated ΔS^0 used.

with the prediction that the fluoride should be less stable than the chloride if $r(AH_4^+) > 1·0$ Å, and the order becomes

$$I > Br > Cl > F.$$

The experimental results therefore suggest that the iodide should be more stable than the bromide if $r(AH_4^+)$ is greater than about 1·5–1·7 Å. The important qualitative feature of the treatment is that the stability order will become $I > Br > Cl > F$ if the radius of the cation is larger than a figure between the values for NH_4^+ and PH_4^+. Hence it is not surprising that experiments involving the condensation of arsine and hydrogen halides on to a film at 110 °K yielded infra-red evidence for the formation of AsH_4^+ in the case of the bromide and iodide only, and suggested that the latter was the more stable.

In the latter part of this chapter, we have seen how a much simplified ionic model is in qualitative accord with the trends in the energies of a number of chemical reactions. This correlation is often observed, even when, for the compounds concerned, the cycle lattice energies and those calculated from (2.7), (2.8) or (2.15) are considerably different, and when there are marked deviations from the properties (see §2.4) normally associated with 'ionic compounds.' Thus one aspect of the ionic model can sometimes serve as a tentative starting point for a qualitative treatment of chemical energies, even when, by the broader criteria discussed in §2.4, the model is not entirely valid.

References and suggestions for further reading

Bills, J. L. and Cotton, F. A. (1960). *J. Phys. Chem.* **64**, 1477.

Brackett, T. E. and Brackett, E. B. (1965). *J. Phys. Chem.* **69**, 3611.

Cubiociotti, D. (1961). *J. Chem. Phys.* **34**, 2189.

Evans, R. C. (1964). *An Introduction to Crystal Chemistry*, 2nd edition, pp. 41–5, 145–150. Cambridge University Press.

Gibb, T. R. (1962). *Progress in Inorganic Chemistry*, Ed. F. Cotton, vol. 3, p. 397. New York: Interscience.

Gray, P. and Waddington, T. C. (1956). *Proc. Roy. Soc.* A **235**, pp. 106, 481.

Kapustinskii, A. F. (1933). *Z. Phys. Chem.* 22B, 257.

Kapustinskii, A. F. (1956). *Quart. Rev. Chem. Soc.* p. 283.

Ladd, M. F. C. and Lee, W. H. 'Progress in Solid State Chemistry', Ed. H. Reiss, vol. I, p. 37 (1964); vol. II, p. 378 (1965). Pergamon Press.

These two review articles contain a very large number of lattice energy values and a good deal of related data.

Morris, D. F. C. (1957). *J. Inorg. Nucl. Chem.* **4**, 8.

Pauling, L. (1960). *The Nature of the Chemical Bond*, 3rd edition, chapter 13. Cornell University Press.

Phillips, C. S. G. and Williams, R. J. P. (1965). *Inorganic Chemistry*, vol. I, pp. 170–5. Oxford University Press.

Pitzer, K. S. and Brewer, L. (1961). *Thermodynamics*, pp. 674–8. New York: McGraw-Hill.

Waddington, T. C. (1959). *Advances in Inorganic Chemistry and Radiochemistry*, Ed. H. J. Emeleus and A. G. Sharpe, vol. I, p. 157, New York: Academic Press. A didactic review article on lattice energies and related topics.

Waddington, T. C. (1966). *Trans. Faraday Soc.* **62**, 1482.

Wells, A. F. (1962). *Structural Inorganic Chemistry*, 3rd edition. Oxford University Press.

the type described in §2.5 indicate that, among the hypothetical dihalides, caesium difluoride is the only one to approach thermo-dynamic stability with respect to the reaction:

$$MX_2(s) \to M(s) + X_2 \text{ (ref. state)}. \tag{3.1}$$

Even in this case, the estimated enthalpy of the decomposition

$$CsF_2(s) \to CsF(s) + \tfrac{1}{2}F_2(g) \tag{3.2}$$

is less than -100 kcal/mole, and the compound would be very unstable with respect to the monofluoride and fluorine.

Similar calculations show that all the alkaline earth mono-halides, save those of beryllium, have negative standard free energies of formation (see reference to Waddington (1959), given on p. 53) but, compared with metals that form stable monohalides, values of I_2 are low and lead to very favourable disproportiona-tion reactions of the type:

$$2MX(s) \to MX_2(s) + M(s). \tag{3.3}$$

As might be expected from the electronic configuration, the third ionization potentials are abnormally large, and compounds of the tripositive oxidation state would presumably be unstable to those in the dipositive one. Thus the thermodynamic evidence suggests that, under normal conditions, the chemistry of the alkali metals and the alkaline earths is confined to that of the elements and one other oxidation state.

Perhaps the most interesting features of this chemistry are the variations in the ease of the decomposition reactions of certain salts from metal to metal. A typical example of the type to be con-sidered in this chapter, is the pyrolysis of the alkaline earth carbonates. This yields the corresponding oxide and CO_2:

$$MCO_3(s) \to MO(s) + CO_2(g) \tag{3.4}$$

and at $25°$, the standard free energies of the reaction vary in the order Mg < Ca < Sr < Ba. We shall first suppose that, for the four carbonates, an equilibrium corresponding to (3.4) is set up in a closed system at all temperatures of interest. Now as an ele-ment makes a roughly constant contribution to the standard entropy of a solid, ΔS^0 will be very nearly the same for the four decomposition reactions, and because

$$(\partial \Delta G^0 / \partial T)p = -\Delta S^0, \tag{3.1}$$

the order of the ΔG^0 values at any temperature is very likely to be the same as at 25°. Under these circumstances, the decomposition temperatures (at which a certain defined equilibrium pressure of carbon dioxide exists over the carbonates) will follow the sequence in ΔG^0_{298} as long as the phases present are correctly described by [3.4]. Finally, as ΔS^0_{298} is nearly constant within these series, the changes in ΔH^0_{298} from metal to metal will in turn be a good guide to those in ΔG^0_{298}.

Although the relation between the sequences in ΔH^0_{298} and the decomposition temperatures can be understood under the conditions that we have specified, it is easy to conceive of circumstances in which it might break down. First, when the compounds or the products of their pyrolyses melt before decomposition, the close similarity in the values of ΔS^0 should disappear at temperatures where some, but not all of the reactants or products in the series of equations are in the fused state. Again, the overall process to which ΔG^0_{298} refers may not represent the equilibrium that is set up during the reaction. Magnesium carbonate for example, after some initial decomposition, forms basic carbonates with the oxide that is produced. The observed decomposition temperature becomes higher because it is a stable phase such as $4MgCO_3 \cdot MgO$ that then dissociates to give the oxide and CO_2. High activation energies may also prevent the equilibrium for the overall decomposition reaction from being reached. Thus the decomposition of potassium perchlorate to potassium chloride and oxygen is favourable at room temperature, so the stability of the compound must be due to a slow rate of reaction. Experimental data on both the formation of solid solutions and other intermediates in solid phase decomposition reactions and on the irreversibility of such reactions at the temperature of decomposition are quite limited. However, as temperatures are raised, reaction rates increase and the inhibiting effects of activation energies should be much diminished.

Despite all these possible objections, there is no doubt that in the great majority of cases, the variations in ΔH^0_{298} or ΔG^0_{298} are an excellent guide to those in the decomposition temperatures of a series of alkali metal or alkaline earth compounds. In the remainder of the chapter this relation is assumed, and the relative thermal stabilities of some series of analogous compounds are interpreted by using (2.14) to explain the variations in the standard enthalpies of the decomposition reactions at 25°. Individual decomposition temperatures quoted in the literature are often unreliable because different workers have very different criteria for what constitutes the onset of decomposition. Where possible, values given in this chapter for a particular series of compounds are those noted by a single observer who used the same criterion throughout. In most cases, they correspond to a particular pressure of a volatile product over the reactant.

3.2. Heats and free energies of formation.
From the Born–Haber cycle for an alkali metal compound MX which is shown in fig. 2.3, (2.1) may be obtained. If M alters and X remains constant, then

$$\Delta H_f^0 = U[MX] - (L_v + I_1) + \text{constant}. \tag{3.2}$$

If the size of the cation is steadily increased, both $U[MX]$ and $(L_v + I_1)$ fall but if, from metal to metal, the stepwise drops in the lattice energy exceed those in the enthalpy of formation of the gaseous cation, then the stabilities of the compounds MX with respect to the constituent elements will decrease with increasing size of the cation. Since the rate of fall of $(L_v + I_1)$ is independent of the nature of the anion, this situation, from (2.14) is most likely to arise when X^- is small. Thus the free energies and enthalpies of decomposition of the alkali metal fluorides become more negative down the series, while those of the iodides become more positive. In the case of the chlorides and bromides, the difference in the rates of decrease of $(L_v + I_1)$ and the lattice energy is insufficient to ensure the dominance of any one term throughout the entire series, and although the overall trend is towards increasing stability with increasing atomic number, slight irregularities are observed (see table 3.3).

Because these compounds boil without decomposition, this argument has little practical relevance to their chemistry, but it is clearly of importance when the stability of the hydrides is considered. On heating, these yield the metal and molecular hydrogen, and the small anion ensures that the drop in the lattice energy is the factor determining the stability variations as the cation radius is raised. Hence at 25°, lithium forms the most stable alkali metal hydride, and while the potassium and sodium analogues evolve

3

TABLE 3.3   *Standard enthalpies of formation of some alkali metal compounds*

$-\Delta H_f^0$ (kcal/mole)

Metal	MF	MCl	MBr	MI	MH	M_3N	M_2O	MOH
Li	146·3	96	83·7	64·8	21·7	47·2	143·4	116·5
Na	136·3	98·2	86·0	68·8	13·6	—	99·4	102·0
K	134·5	104·2	93·7	78·3	15·2	—	86·4	101·8
Rb	131·8	103·4	93·5	79·0	11·4	—	78·9	98·9
Cs	130·3	106·9	97·7	83·9	10·1	—	75·9	97·2

hydrogen at about 400°, the lithium compound melts at nearly 700° without decomposition. Consequently lithium metal may be purified of traces of potassium by heating it in hydrogen to 800°. At this temperature, potassium distils off while lithium remains as the pure molten hydride which can be decomposed by further heating to 1000°.

Similar stability orders are observed with other compounds of small anions. Lithium is the only alkali metal to combine directly with molecular nitrogen. Active nitrogen in an electric discharge at 0° converts sodium first to the nitride Na_3N, but further action yields the azide and potassium and rubidium do not even form the nitride as an intermediate. The question of the stability of the nitrides and nitrogen with respect to the azides can be successfully treated by a cycle of the type discussed in §3.4. If, however, we assume that the formation of a nitride is controlled by the thermo-dynamic stability with respect to the elements in their standard states, then on the ionic model implied by the simplest Kapustin-skii equation, the extreme gradation in stability of the nitrides is due to the high charge and small size of the anion: compared with a compound of a singly charged anion of equal size, the lattice energy of the nitride M_3N falls six times (as $z_-[z_- + 1]$) as rapidly, while the stepwise drops in the ionization and sublimation enthalpy terms are only three times as great. An increase in the charge of the anion thus favours the phenomenon of diminishing stability with respect to the elements down the series.

This point is brought out more clearly in the alkaline earth group for which figures are given in table 3.4. Examination of the

TABLE 3.4 *Standard enthalpies of formation of some alkaline earth compounds*

$-\Delta H_f^0$ (kcal/mole)

Metal	MF_2	MH_2	MCl_2	MBr_2	MI_2	MO	M_3N_2
Mg	264·9	17·2	153·2	123·7	86·0	143·8	116·0
Ca	290·2	41·7	190·0	161·3	127·8	151·8	103·2
Sr	290·3	42·3	198·0	171·1	135·5	141·1	93·4
Ba	286·9	40·9	205·6	180·4	144·0	133·5	86·9

standard enthalpies and free energies of formation of compounds shows that, even with the smallest singly and doubly charged anions, the fall in the sum of the latent heat of sublimation and the ionization enthalpies between magnesium and calcium is large enough to exceed in every known case the drop in the lattice energy. Thus for these compounds, the usual order of free energies of formation is Mg > Ca ~ Sr < Ba. MgH_2, which has the rutile structure, is much the least stable of the four hydrides. It must be prepared by heating the metal in 200 atmospheres of the gas, and it begins to lose hydrogen at about 300°, some 300° below the decomposition temperatures of the corresponding calcium, strontium and barium compounds. This contrasts with the behaviour of lithium hydride. Only in the case of the small triply charged nitride anions is the sequence Mg < Ca < Sr < Ba observed. This last order of stability is, as expected, completely reversed for the medium and large sized singly charged anions.

3.3. Decomposition of compounds into products containing anions of the same charge. An alkali metal polyiodide, MI_3, tends to be unstable with respect to the iodide and iodine. The variations in the standard free energy changes of this decomposition may be discussed in terms of the thermodynamic cycle shown in fig. 3.1.

$$\Delta H^0 = U[MI_3] - U[MI] + x. \tag{3.3}$$

From (2.14),

$$\Delta H^0 = 512 \left[\frac{1}{r(M^+) + r(I_3^-)} - \frac{1}{r(M^+) + r(I^-)} \right] + x, \tag{3.4}$$

$$\frac{d(\Delta H^0)}{dr(M^+)} = -512 \left[\frac{1}{[r(M^+) + r(I_3^-)]^2} - \frac{1}{[r(M^+) + r(I^-)]^2} \right]. \tag{3.5}$$

The numerators in the Kapustinskii expressions for the lattice energies are the same for both the simple and complex iodide, and we can safely assume that the effective thermochemical radius of the anion in the former is less than that in the latter. Hence, from (3.4), the stepwise drops in the lattice energy of the iodide will exceed those in the lattice energy of the polyiodide as the cation radius is raised and, from lithium to caesium, ΔH^0 should become more positive. A more mathematical representation of the argument is shown in (3.5). $d(\Delta H^0)/dr(M^+)$ is clearly positive if $r(I_3^-) > r(I^-)$.

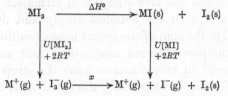

Fig. 3.1. Thermodynamic cycle for the decomposition of an alkali metal tri-iodide.

The ionic model therefore accounts successfully for the observation that, in a particular series of alkali metal polyhalides, the stability increases from lithium to caesium. In the tri-iodide series for example, only rubidium and caesium form anhydrous solids, and the latter is the more stable. Stable compounds of the large tetraalkylammonium ions like $N(CH_3)_4^+$ and $N[C(CH_3)_3]_4^+$ are also known. Recently the compounds $KClF_2$, $RbClF_2$ and $CsClF_2$ were prepared by heating the fluorides and ClF to $175°$ under pressure. Their respective decomposition temperatures were $237°$, $248°$ and $262°$ and attempts to prepare the lithium and sodium salts were unsuccessful. Again the dissolution of the alkali metal fluorides in liquid ClF_3 and BrF_5 at $100°$, followed by subsequent evaporation, yields products containing the compounds $KClF_4$, $RbClF_4$ and $CsClF_4$ and $KBrF_6$, $RbBrF_6$ and $CsBrF_6$. No lithium or sodium salts could be prepared. $CsClF_4$ was stable to $300°$ but $KClF_4$ began decomposing at $200°$.

This argument can be extended very easily to account for what is, relative to the other alkali metals, the ready decomposition of lithium carbonate to the oxide and CO_2, and for the marked instability of lithium peroxide which begins to lose oxygen at

TABLE 3.5 *Some thermodynamic properties (kcal/mole) and decomposition temperatures (°C) for the pyrolyses of some alkali metal and alkaline earth compounds*

$$MCO_3(s) \rightarrow MO(g) + CO_2(g)$$

	Mg	Ca	Sr	Ba
ΔG^0_{298}	15·6	31·2	43·8	51·5
$t^{(b)}$	400	900	1280	1360

$$M(OH)_2(s) \rightarrow MO(s) + H_2O(g)$$

	Mg	Ca	Sr	Ba
ΔG^0_{298}	8·5	15·4	19·4	23·7
$t^{(b)}$	300	390	466	700

$$M(NO_3)_2(s) \rightarrow MO(s) + 2NO(g) + \tfrac{3}{2}O_2(g)^{(a)}$$

	Be	Mg	Ca	Sr	Ba	Ra
ΔG^0_{298}	—	45·9	74·4	93·6	105·0	114·2
$t^{(b)}$	125	450	575	635	675	—

$$MO_2(s) \rightarrow MO(s) + \tfrac{1}{2}O_2(g)$$

	Mg	Ca	Sr	Ba
ΔH^0_{298}	5·1	5·7	12·5	17·0
$t^{(b)}$	375	380	480	790

$$MO_2(s) \rightarrow \tfrac{1}{2}M_2O_2(s) + \tfrac{1}{2}O_2(g)^{(a)}$$

	Li	Na	K	Rb	Cs
ΔH^0_{298}	LiO$_2$	1·6	8	12·3	14·0
$t^{(b)}$	Unknown	100	471	600	900

$$MHF_2(s) \rightarrow MF(s) + HF(g)$$

	Li	Na	K	Rb	Cs
ΔH^0_{298}		15·5	20·6	21·2	24·4
$t^{(b)}$	< 200°	278°	400°	> 400°	> 400°

(a) Precise stoicheiometry of the decomposition reaction is rather uncertain.
(b) Temperature required to reach a pressure of the gaseous products that is constant for the series.

about 300°, the temperature at which Na_2O_2 is commercially prepared by heating the metal in air. It is also in accord with the standard free energies and temperatures of decomposition of the alkali metal bifluorides and alkaline earth carbonates and peroxides shown in table 3.5. The high stability of barium peroxide was formerly used for the commercial preparation of oxygen. Barium oxide was heated in air to 500° to increase the rate of oxygen uptake, and the resulting peroxide then decomposed by raising the temperature to 800°.

3.4. Decomposition of compounds into products containing anions of different charge.

By a slight modification of the arguments in the previous section, the variations in the stabilities of the alkali metal superoxides may be understood. Considering decomposition into the peroxides, the cycle in fig. 3.2 yields the relation:

$$\Delta H^0 = 2U[MO_2] - U[M_2O_2] + RT + x. \tag{3.6}$$

Using (2.14),

$$\Delta H^0 = 256 \left[\frac{4}{r(M^+) + r(O_2^-)} - \frac{6}{r(M^+) + r(O_2^{2-})} \right] + RT + x, \tag{3.7}$$

$$\therefore \frac{d(\Delta H^0)}{dr(M^+)} = -256 \left[\frac{4}{[r(M^+) + r(O_2^-)]^2} - \frac{6}{[r(M^+) + r(O_2^{2-})]^2} \right]. \tag{3.8}$$

The bond lengths in the peroxide and superoxide ions in solid compounds differ by only 0·2 Å, and we shall assume that their thermochemical radii are the same. Thus the sign of $d(\Delta H^0)/dr(M^+)$ is determined by the numerators in (3.8), and the stability of the superoxides with respect to the peroxides (and monoxides) should increase down the alkali metal series. As the data in table 3.5 indicate, this is indeed the case. The combination of radius and charge effects accounts for the variations in the stabilities of the various oxides of the alkali metals. Lithium peroxide, as discussed in §3.3, is less stable than sodium peroxide. At potassium the superoxide is stable with respect to the peroxide and the decomposition becomes progressively less favourable at rubidium and caesium. This sequence is reflected in the oxides formed by the combustion of the metals in oxygen: Li_2O, Na_2O_2, KO_2, RbO_2 and

CsO_2. Commercial sodium peroxide, which is made by oxidation of the metal, contains about 10 per cent of the yellow superoxide.

Table 3.5 also includes data on the decomposition of the alkaline earth hydroxides. This problem is another in which the anions in the reactant and product have similar sizes. It may be discussed by a cycle which is very similar to that in fig. 3.2.

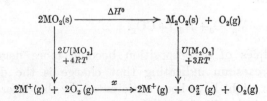

Fig. 3.2. Thermodynamic cycle for the decomposition of an alkali metal superoxide.

When substantial radius and charge effects act together, the changes in the feasibility of a particular reaction as the cation alters may be quite violent. In the case of the alkali metal nitrates, only the lithium compound yields the oxide on heating:

$$2MNO_3 \to M_2O + 2NO + \tfrac{3}{2}O_2. \tag{3.5}$$

By using a cycle of the type shown in fig. 3.2, we obtain:

$$\Delta H^0 = 2U[MNO_3] - U[M_2O] + c \tag{3.9}$$

$$= 256 \left[\frac{4}{r(M^+)+r(NO_3^-)} - \frac{6}{r(M^+)+r(O^{2-})} \right] + c, \tag{3.10}$$

where c is a constant as the cation varies. In this case, the numerator in the lattice energy term for the oxide is larger than that in the corresponding term for the nitrate while the denominator is less. This leads to an unusually rapid increase in stability with cation size, while the decomposition into the nitrite embodies only a small radius effect. The latter reaction diminishes slightly in feasibility as one descends the series and is overtaken by the rapidly increasing free energy change associated with the monoxide decomposition. Thus the nitrates of sodium, potassium, rubidium and caesium yield the nitrite when heated to moderate temperatures. In contrast, under similar conditions the oxide is the main product of the pyrolysis of all the alkaline earth nitrates. As table 3.5 shows, the decomposition temperatures and values of ΔG^0 are in the expected order.

When they are in competition with one another, either radius or charge effects may dominate the stability sequence. Charge effects are more influential in the superoxide decomposition discussed earlier in this section and in the pyrolysis of the alkali metal bicarbonates:

$$2\text{MHCO}_3 \to \text{M}_2\text{CO}_3 + \text{H}_2\text{O} + \text{CO}_2. \tag{3.6}$$

The enthalpies of decomposition become more negative from sodium to caesium indicating that charge is the determining factor. This might have been anticipated for the difference in the thermochemical radii of the carbonate and bicarbonate ions is only about 0·2 Å, the latter being the smaller. Presumably then, the decomposition becomes slightly less favourable as the series is descended, and because the bicarbonates lie close to the borderline of stability at room temperature and pressure, it is not surprising to find that the lithium compound has not yet been prepared. The low solubility of lithium carbonate is another obstacle to the preparation of LiHCO_3. This removes carbonate ions from the aqueous solution and swings the equilibrium:

$$2\text{HCO}_3^- \to \text{H}_2\text{O} + \text{CO}_2 + \text{CO}_3^{2-} \tag{3.7}$$

to the right.

From solutions of alkali metal fluorides in liquid xenon hexafluoride, the compounds RbXeF_7 and CsXeF_7 may be obtained, but the lithium, sodium and potassium and analogues are unstable to the reaction,

$$2\text{MXeF}_7 \to \text{M}_2\text{XeF}_8 + \text{XeF}_6. \tag{3.8}$$

Furthermore, the standard enthalpies of decomposition of the rubidium and caesium salts are $+9$ and $+14$ kcal/mole. Using a restatement of the type shown in fig. 3.2, we will obtain after differentiation,

$$\frac{\mathrm{d}(\Delta H^0)}{\mathrm{d}r(\text{M}^+)} = -256 \left[\frac{4}{[r(\text{M}^+) + r(\text{XeF}_7^-)]^2} - \frac{6}{[r(\text{M}^+) + r(\text{XeF}_8^{2-})]^2} \right].$$

$$\tag{3.11}$$

Since the difference in the sizes of the two anions is likely to be relatively small, it is not surprising to find that charge effects dominate and that the stability sequence should be Li, Na, K < Rb < Cs. If the rubidium and caesium salts are warmed, they decompose at 0° and 50° respectively in accordance with [3.8], while from the original solutions of the alkali metal fluorides in XeF_6, some evidence for the formation of Na_2XeF_8 and K_2XeF_8 but not Li_2XeF_8 has been obtained. The four M_2XeF_8 compounds are thought to be unstable to reactions of the type,

$$M_2XeF_8 \rightarrow 2MF + XeF_6, \qquad [3.9]$$

and the sodium and potassium compounds decompose, sometimes explosively, at temperatures well below 250°, while the evolution of gases from the caesium and rubidium analogues does not begin until a temperature of about 400° is attained. For [3.9] we may write,

$$\frac{d(\Delta H^0)}{dr(M^+)} = -256 \left[\frac{6}{[r(M^+) + r(XeF_8^{2-})]^2} - \frac{4}{[r(M^+) + r(F^-)]^2} \right] \quad (3.12)$$

and it now appears that the difference in the radii of the anions is so large that charge effects are swamped and the stability sequence again becomes Li < Na, K < Rb, Cs. In this and other series of salts containing large anions, such a sequence is favoured by the destabilizing effect of anion–anion contact in compounds of the smaller cations.

3.5. The anomalous nature of lithium. The evidence for the 'anomalous' nature of lithium compounds and their similarities to magnesium analogues rests upon features like the thermal decomposition of the nitrate and carbonate with formation of the monoxide, and the production of a nitride and hydride by direct combination of the heated elements. The content of the three preceding sections indicates that the behaviour of lithium in all these respects fits into the expected pattern of gradual changes that occur in the chemistry of a series of alkali metal salts as the size of the cation is decreased. It is true that the chemical changes that occur when we pass from lithium to sodium are more violent than those that are observed between any other pair of adjacent alkali metals, but speaking thermodynamically, we can attribute this to the large percentage change in ionic radius between the metals of the second and third short periods, and partly to the chance conditions of temperature and pressure which commonly prevail.

If we consider the alkali metal carbonates, the lattice energy falls by over 10 per cent between lithium and sodium and between sodium and potassium, but only by about 5 per cent between the remaining members of the series. When this effect is combined with the diminishing *absolute* values of the lattice energies, it is clear that larger changes in the thermodynamics of analogous reactions will occur between lithium and sodium and sodium and potassium than between any other pair of adjacent elements in the series. Again, the observation that a compound decomposes in the range 0–900° under an arbitrary pressure of the gaseous products while another similar compound does not, indicates no fundamental difference between them. Indeed, in the case of the carbonates of lithium and sodium it results from a difference of 13 kcal/mole in the standard free energies of decomposition, the separate values being 43 and 66 kcal/mole respectively. The differences between lithium and potassium, the typical alkali metal, are therefore readily understood in terms of an ionic model extended to small values of the cation radius, and the unusual solubilities of certain lithium salts are similarly covered in chapter 5. When these facts are appreciated, the 'anomalous' nature of lithium is seen in its proper perspective.

Since the carbonate and nitrate of lithium are less stable than the corresponding compounds of the other alkali metals, it is not surprising that they should resemble their magnesium analogues in yielding the monoxide on heating. When the close similarity in the ionic radii of Li^+ and Mg^{2+} is allied with the ionic model, the relative ease of these decompositions is obtained.

If the enthalpies of decomposition of lithium and magnesium carbonates are denoted by ΔH_1^0 and ΔH_2^0, then a thermodynamic cycle of the type shown in fig. 3.1 yields the following equations:

$$\Delta H_1^0 = U[Li_2CO_3] - U[Li_2O] + x, \tag{3.13}$$

$$\Delta H_2^0 = U[MgCO_3] - U[MgO] + x. \tag{3.14}$$

The denominators in the Kapustinskii lattice energy terms for the two carbonates and the two oxides are virtually the same in both equations, but the numerators for the compounds of lithium are only $\frac{6}{8}$ those of magnesium. As the constant x is the same in both equations and the lattice energies of the oxides exceed those of the corresponding carbonates, ΔH_2^0 is more negative than ΔH_1^0 and the carbonate or nitrate of lithium should be more stable than the carbonate or nitrate of magnesium. Magnesium carbonate begins to lose carbon dioxide at about 550° while lithium carbonate is stable up to 700°. The corresponding values for the nitrates are 300° and 500° respectively.

These decompositions are specific examples of a fairly general principle. Because the alkaline earth ions are smaller than their alkali metal neighbours, and because, in decomposition reactions, the anions in the products are usually smaller than those in the reactants, the alkaline earths are generally poorer anion stabilizers than the alkali metals. A polyhalide of barium, for example, is nearly always less stable than its potassium, rubidium or caesium analogue. Another contributory factor may be the destabilizing effect of anion–anion contact which would be more significant in alkaline earth compounds where the proportion of anions is higher.

In conclusion, we emphasise that the successful applications of equation (2.14) described in §2.8–3.5 are in a sense surprising in view of the approximations made in deriving the Kapustinskii equation (see §2.7) and possible breakdowns in the correlation between decomposition temperature and ΔG_{298}^0 (see §3.1). More accurate information may well expose these inadequacies. Above all, our success is in no sense an indication that the compounds concerned would be regarded as 'ionic' by more searching criteria (see §2.4 and page 52). We note too, that the relation between cation radius and the stability of compounds to dissociation cannot be extended to embrace the 'B metals' or the transition elements. Thus, although the nickel and magnesium, and the cadmium and calcium cations are of comparable size, the alkaline earth carbonate is considerably more stable in each case. Contributions to the lattice energies of the nickel and cadmium oxides which are not covered by equations like (2.7) or (2.14), stabilize the compounds and enhance dissociation.

4. Solution Equilibria and Electrode Potentials

4.1. The enthalpies and free energies of formation of aquated ions. The formation of an ion from its constituent elements is most readily represented by equations of the type:

$$Cd(s) \rightarrow Cd^{2+}(aq) + 2e \qquad [4.1]$$

$$Mn(s) + 2O_2(g) + e \rightarrow MnO_4^-(aq). \qquad [4.2]$$

Unfortunately, solutions containing only one type of ion are unobtainable, so that experimental measurement of the associated energy changes is impossible, and there is the additional problem of assigning an arbitrarily chosen state to the electron in all reactions of this kind. Both difficulties can be removed by defining the thermodynamics of formation of ions at any temperature relative to those associated with the production of the appropriate number of aquated protons from molecular hydrogen at that temperature:

$$Cd(s) + 2H^+(aq) \rightarrow Cd^{2+}(aq) + H_2(g) \qquad [4.3]$$

$$Mn(s) + \tfrac{1}{2}H_2(g) + 2O_2(g) \rightarrow MnO_4^-(aq) + H^+(aq). \qquad [4.4]$$

An indirect method is necessary to obtain ΔG^0 and ΔH^0 for [4.4], but the thermodynamic changes for both the above equations are capable of experimental measurement. It is these quantities that are tabulated as ΔG_f^0 and ΔH_f^0 of the appropriate ions in compilations of thermodynamic properties. The standard state for these ions is that defined in appendix 1. This definition of the standard free energy and enthalpy of formation of ions is equivalent to a new convention assigning zero ΔH_f^0 and ΔG_f^0 to the aquated proton at all temperatures, for these properties are then defined as those associated with the isothermal reaction:

$$\tfrac{1}{2}H_2(g) + H^+(aq) \rightarrow H^+(aq) + \tfrac{1}{2}H_2(g). \qquad [4.5]$$

[68]

In any isothermal reaction with balanced charges involving aqueous ions, the relations:

$$\Delta H^0 = \Sigma \Delta H_f^0 \text{ (products)} - \Sigma \Delta H_f^0 \text{ (reactants)} \tag{4.1}$$

and

$$\Delta G^0 = \Sigma \Delta G_f^0 \text{ (products)} - \Sigma \Delta G_f^0 \text{ (reactants)} \tag{4.2}$$

now hold.

4.2. The entropies of aquated ions.

In a saturated solution of sodium chloride in contact with the crystalline solid, there exists the equilibrium:

$$\text{NaCl(s)} \to \text{Na}^+(\text{aq}) + \text{Cl}^-(\text{aq}) \tag{4.6}$$

for which

$$\Delta G^0 = -RT \ln a_{\text{Na}^+} a_{\text{Cl}^-}, \tag{4.3}$$

$$= -RT \ln m_{\text{Na}^+} \gamma_{\text{Na}^+} m_{\text{Cl}^-} \gamma_{\text{Cl}^-}, \tag{4.4}$$

$$= -RT \ln m^2 \gamma_{\pm}^2, \tag{4.5}$$

where the symbols a and m with ionic subscripts are the activities and molalities of the corresponding ions, m is the molality of sodium chloride and $\gamma_{\pm}$ is the mean ion activity coefficient in the saturated solution. Thus solubility measurements and activity coefficient data can be used to find the standard free energy of solution of sodium chloride and this, when combined with a direct measurement of the heat of solution, will give the standard entropy of [4.6]. Since the standard molal entropy of sodium chloride at 25° can be found by specific heat measurements, the sum of the standard molal entropies of the sodium and chloride ions is readily obtained, but to assign individual entropies another convention must be introduced, for while the experimental measurement of ΔS^0 for ionic reactions is possible, no such reaction contains only one ion. It is customary to assign a standard molal entropy of zero to the aquated proton at all temperatures. Then, for example, ΔS^0 for [4.3], together with the entropies of hydrogen gas and cadmium metal, yields the molal entropy of aquated cadmium ions. With this convention, in any isothermal ionic reaction with balanced charges,

$$\Delta S^0 = \Sigma S^0 \text{ (products)} - \Sigma S^0 \text{ (reactants)}. \tag{4.6}$$

Some relevant thermodynamic data is given in table 4.1.

When an ion is transferred from the gaseous to the aqueous phase, its standard molal entropy is markedly decreased. This decrease, which is numerically larger when the ion is smaller or more highly charged, may be correlated with the restriction imposed in solution on the movements of the adjacent, polar water molecules by interaction with the ions, although detailed treatments suggest that a number of other effects should be considered, and that the influence of dissolved electrolytes on the structure of water is rather complex (Hunt, 1963).

TABLE 4.1   *Thermodynamic properties of liquid water and some aquated ions at 25°*

Formula	ΔH_f^0	ΔG_f^0	S^0
H_2O	$-68 \cdot 32$	$-56 \cdot 69$	$16 \cdot 71$
H^+	0	0	0
OH^-	$-54 \cdot 97$	$-37 \cdot 59$	$-2 \cdot 57$
Na^+	$-57 \cdot 28$	$-62 \cdot 59$	$14 \cdot 4$
Cd^{2+}	$-17 \cdot 3$	$-18 \cdot 58$	$-14 \cdot 6$
Cu^{2+}	$15 \cdot 4$	$15 \cdot 5$	$-23 \cdot 6$
Fe^{2+}	$-21 \cdot 0$	$-20 \cdot 3$	$-27 \cdot 1$
Fe^{3+}	$-11 \cdot 4$	$-2 \cdot 5$	$-70 \cdot 1$
Pu^{4+}	$-129 \cdot 1$	$-118 \cdot 2$	-87
Cl^-	$-39 \cdot 95$	$-31 \cdot 37$	$13 \cdot 5$
S^{2-}	$7 \cdot 9$	$20 \cdot 5$	$-3 \cdot 5$
MnO_4^-	$-129 \cdot 7$	$-107 \cdot 1$	$45 \cdot 4$
SO_4^{2-}	$-217 \cdot 3$	$-178 \cdot 0$	$4 \cdot 8$
PO_4^{3-}	$-305 \cdot 9$	$-244 \cdot 0$	-53

Under the same conditions of temperature and pressure, the variations in the entropies of gaseous, monatomic ions are almost entirely determined by a term, $(\frac{3}{2}R \ln M + R \ln Q)$, in the Sackur–Tetrode equation, where M is the atomic weight and Q is the electronic multiplicity of the ground state. Throughout the entire range of known monatomic ions, the changes in this term are comparatively slight, so that the variations in the entropies of ions in water are largely determined by those in the entropy change associated with the transfer from the gaseous to the aqueous phase. Consequently, as the data in table 4.1 shows, the standard molal entropies of ions diminish with increasing charge and decreasing size. Latimer and Powell (1951) placed this result on a quantitative basis by means of an empirical relation. If the molal entropies of ions are related to that of the aquated proton as zero, then, for monatomic ions, the following equation is approximately correct:

$$S^0 = \tfrac{3}{2}R \ln M - \frac{270z}{(r+x)^2} + 37. \tag{4.7}$$

Here z is the charge on the ion, r is Pauling's crystal radius in Angstrom units, x is a constant with the value 1.00 Å for anions and 2.00 Å for cations, and M is the atomic weight.

4.3. Manipulation of the thermodynamic properties of ions. For permanganate ion, the relation between the three quantities listed in table 4.1 is as follows: ΔG_f^0, ΔH_f^0 and ΔS_f^0 refer to [4.4] and

$$\Delta S_f^0 = S^0[\mathrm{MnO_4^-}\,(\mathrm{aq})] - (S^0[\mathrm{Mn}\,(\mathrm{s})] + \tfrac{1}{2}S^0[\mathrm{H_2}\,(\mathrm{g})] + 2S^0[\mathrm{O_2}\,(\mathrm{g})]), \tag{4.8}$$

$$= -75.8 \text{ cal/deg.mole.} \tag{4.9}$$

Then, at $25°$, as expected:

$$\Delta G_f^0 = \Delta H_f^0 - T\Delta S_f^0 = -107.1 \text{ kcal/g-ion.} \tag{4.10}$$

Again, for a monatomic cation M^{n+}, at $25°$:

$$\Delta G_f^0 = \Delta H_f^0 - 0.298(S^0[M^{n+}\,(\mathrm{aq})] + 15.6n - S^0[M\,(\mathrm{s})]). \tag{4.11}$$

4.4. The e.m.f. of cells. Fig. 4.1 shows a rod of copper in a solution of cupric ions connected by a conducting salt bridge containing potassium chloride solution to a hydrogen electrode. The latter consists of a piece of platinum foil coated electrolytically with platinum black and partially immersed in a solution of hydrogen ions. Hydrogen is passed through the electrode in such a way that the platinum is in contact with both the solution and the gas. By the absorption of hydrogen, probably with the formation of surface hydrides, the platinum lowers the activation energy associated with the conversion of hydrogen to hydrogen ions and the electrode becomes reversible.

If a potentiometer is now connected across the poles of the

electrode, a potential difference is observed between them. If the potentiometer can be adjusted until the current is zero and the tests for reversibility at the balance point are successful, then the reading on the instrument gives the numerical value of the e.m.f., E, of the cell. When the electrodes are joined by a conducting wire rather than a potentiometer then a current flows, hydrogen dissolves to give hydrogen ions and fresh copper is deposited on the

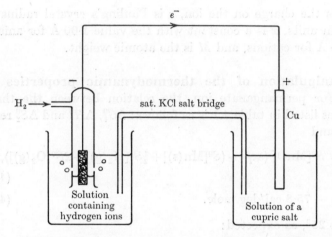

Fig. 4.1. The cell, $^-Pt, H_2(g)|H^+(aq) \vdots KCl (sat.) \vdots Cu^{2+}(aq)|Cu^+$.

metal rod. The pole towards which the electrons move *in the external circuit* is called the positive pole and the sign of the e.m.f. of the cell is defined as that of the pole on the right, i.e. it depends only on the way that the cell is written down. Thus in fig. 4.1, the e.m.f. is positive but if the hydrogen and copper electrodes were interchanged to form the cell:

$$^+Cu|Cu^{2+}(aq) \vdots KCl \vdots H^+(aq)|H_2,Pt^-,$$

then by definition the sign of the e.m.f. would be negative even though the chemical behaviour of the cell was unchanged. When the cell reaction is written such that oxidation occurs at the left-hand electrode and reduction at the right, it can be shown that the reversible e.m.f. of the cell with the defined sign is related to the molal free energy change by the equation:

$$\Delta G = -nFE. \tag{4.12}$$

Here n is the number of electrons released at the reducing electrode when the number of molecules of the reductant involved in the cell equation are oxidized.

Writing the reaction for the cell in fig. 4.1 as,

$$Cu^{2+}(aq) + H_2(g) \rightarrow Cu(s) + 2H^+(aq) \qquad [4.7]$$

then $n = 2$ and by the reaction isotherm:

$$-\Delta G = -\Delta G^0 + RT \ln \frac{a_{Cu^{2+}} \cdot a_{H_2}}{a_{Cu} \cdot a_{H^+}^2}, \qquad (4.13)$$

$$\therefore E = E^0 + \frac{RT}{2F} \ln \frac{a_{Cu^{2+}} a_{H_2}}{a_{H^+}^2}. \qquad (4.14)$$

In a 1·18M solution of hydrogen chloride, the mean ion activity is unity, so if the hydrogen electrode is made up with this solution and hydrogen gas at one atmosphere fugacity:

$$E = E^0 + \frac{RT}{2F} \ln a_{Cu^{2+}}, \qquad (4.15)$$

$$= E^0 + \frac{RT}{2F} \ln m_{Cu^{2+}} + \frac{RT}{2F} \ln \gamma_{Cu^{2+}}. \qquad (4.16)$$

At infinite dilution, $\gamma_{Cu^{2+}} = 1$, so by plotting $E - \dfrac{RT}{2F} \ln m_{Cu^{2+}}$ against a suitable function of the molality (usually $\sqrt{m}$) and extrapolating to $m = 0$, E^0 is obtained.

Since $\Delta G^0 = -\Delta G_f^0[Cu^{2+}(aq)]$, this experiment affords a convenient method of measuring not only the standard e.m.f. of the cell, but also the standard free energy of formation of aquated cupric ions.

Standard free energies of many other reactions of the type:

$$Reactants + \tfrac{1}{2}n \, H_2(g) \rightarrow nH^+(aq) + products \qquad [4.8]$$

may be measured by electrical methods. The cells

$$^-Pt, H_2 | HClO_4, Fe^{2+}, Fe^{3+} | Pt^+$$

and

$$^-Pt, H_2 | HClO_4, Mn(ClO_4)_2 | \beta - MnO_2, Pt^+$$

contain examples of more complicated electrodes. Their standard e.m.f.s, at unit activity of hydrogen ions, are found to be $+0·77$ V and $+1·23$ V respectively.

In practice the hydrogen electrode is often replaced by a more convenient system like the saturated calomel electrode which has been previously calibrated against it.

Quite frequently the attempted measurement is useless because the system turns out to be irreversible, i.e. thermodynamic equilibrium is not attained in the cell. It is then necessary to use less direct methods which effectively involve the individual determination of the standard free energies of formation of reactants and products.

4.5. Electrode potentials. A series of standard e.m.f.'s of cells whose chemical processes are of the type shown in [**4.8**] are recorded in table 4.2. It will be noted that all the equations include a term ne on the left-hand side where n is an integer. This symbol merely indicates that $\frac{1}{2}nH_2(g)$ should be added to the left-hand side and $nH^+(aq)$ to the right, both hydrogen and hydrogen ions being at unit activity. To take it at the literal value would be meaningless because the thermodynamic state of the electron is unspecified, and even if this omission were corrected, the E^0 value of the reaction would be very different from that given in table 4.2, unless the assigned state were a highly artificial one.

On the recommendation of the International Union of Pure and Applied Chemistry (IUPAC), only the potentials with the signs given in table 4.2 are known as standard electrode potentials. In this book, the corresponding cell reaction has been written with the term $\frac{1}{2}nH_2$ on the left and nH^+ on the right, for only then is the equation:

$$\Delta G^0 = -nFE^0 \tag{4.17}$$

satisfied, E^0 being the standard electrode potential.

Some American textbooks ignore the IUPAC recommendation and present electrode potentials with the opposite sign. Then, if (4.17) is to be satisfied, the cell reaction must be written with the term nH^+ on the left and $\frac{1}{2}nH_2$ on the right:

$$Fe^{2+} + H^+ \rightarrow Fe^{3+} + \tfrac{1}{2}H_2, \quad E^0 = -0.77 \text{ V} \tag{4.9}$$

or

$$Fe^{2+} \rightarrow Fe^{3+} + e \text{ (abbreviated).} \tag{4.10}$$

TABLE 4.2 *Some electrode potentials in aqueous solution at 25°*

Electrode	E^0 (Volts)[a]
$Cs^+ + e \rightarrow Cs$	$-3 \cdot 07$
$Li^+ + e \rightarrow Li$	$-3 \cdot 05$
$K^+ + e \rightarrow K$	$-2 \cdot 93$
$Rb^+ + e \rightarrow Rb$	$-2 \cdot 93$
$Ba^{2+} + 2e \rightarrow Ba$	$-2 \cdot 90$
$Sr^{2+} + 2e \rightarrow Sr$	$-2 \cdot 89$
$Ca^{2+} + 2e \rightarrow Ca$	$-2 \cdot 87$
$Na^+ + e \rightarrow Na$	$-2 \cdot 71$
$Mg^{2+} + 2e \rightarrow Mg$	$-2 \cdot 37$
$Zn^{2+} + 2e \rightarrow Zn$	$-0 \cdot 76$
$Cr^{3+} + e \rightarrow Cr^{2+}$	$-0 \cdot 41$
$V^{0+} + e \rightarrow V^{2+}$	$-0 \cdot 26$
$H^+ + e \rightarrow \frac{1}{2}H_2$	0
$Ti^{IV} + e \rightarrow Ti^{3+}$	$0 \cdot 1$[b]
$Sn^{4+} + 2e \rightarrow Sn^{2+}$	$0 \cdot 15$
$Cu^{2+} + 2e \rightarrow Cu$	$0 \cdot 34$
$Fe(CN)_6^{3-} + e \rightarrow Fe(CN)_6^{4-}$	$0 \cdot 36$
$\frac{1}{2}I_2 + e \rightarrow I^-$	$0 \cdot 54$
$Fe^{3+} + e \rightarrow Fe^{2+}$	$0 \cdot 77$
$Ag^+ + e \rightarrow Ag$	$0 \cdot 80$
$Hg^{2+} + 2e \rightarrow Hg$	$0 \cdot 85$
$\frac{1}{2}Br_2 + e \rightarrow Br^-$	$1 \cdot 07$
$O_2 + 4H^+ + 4e \rightarrow 2H_2O$	$1 \cdot 23$
$MnO_2 + 4H^+ + 2e \rightarrow Mn^{2+} + 2H_2O$	$1 \cdot 23$
$Cr_2O_7^{2-} + 14H^+ + 6e \rightarrow 2Cr^{3+} + 7H_2O$	$1 \cdot 33$
$\frac{1}{2}Cl_2 + e \rightarrow Cl^-$	$1 \cdot 36$
$ClO_4^- + 8H^+ + 8e \rightarrow Cl^- + 4H_2O$	$1 \cdot 39$
$MnO_4^- + 8H^+ + 5e \rightarrow Mn^{2+} + 4H_2O$	$1 \cdot 51$
$BrO_3^- + 6H^+ + 5e \rightarrow \frac{1}{2}Br_2 + 3H_2O$	$1 \cdot 52$
$Co^{9+} + e \rightarrow Co^{2+}$	$1 \cdot 95$
$Ag^{2+} + e \rightarrow Ag^+$	$1 \cdot 98$
$S_2O_8^{2-} + 2e \rightarrow 2SO_4^{2-}$	$2 \cdot 01$
$\frac{1}{2}F_2 + e \rightarrow F^-$	$2 \cdot 87$

[a] Values mainly from Latimer (1952).
[b] The nature of Ti^{IV} in solution is rather uncertain.

With the convention used in this book, the couples containing the powerful reducing agents (Na, Cr^{2+}) have negative potentials, while those with strong oxidizing agents (MnO_4^-, $S_2O_8^{2-}$) take positive values. For a reaction of the type:

$$aA + bB + cC + ne \rightarrow kK + lL + mM \qquad [4.11]$$

the variation of the electrode potential with the activities of reactants and products is given by:

$$E = E^0 + \frac{RT}{nF} \ln \frac{a_A^a \cdot a_B^b \cdot a_C^c}{a_K^k \cdot a_L^l \cdot a_M^m}. \tag{4.18}$$

When the potentials have the opposite sign [4.11] should be reversed to fulfil (4.12) and (4.17). Powerful reducing couples then have positive potentials and strong oxidizing couples have negative ones. Under these circumstances,

$$E = E^0 - \frac{RT}{nF} \ln \frac{a_A^a \cdot a_B^b \cdot a_C^c}{a_K^k \cdot a_L^l \cdot a_M^m}. \tag{4.19}$$

In (4.18) and (4.19), values of E still refer to the *standard* hydrogen electrode. Converting to logarithms to the base 10 at 25° and substituting constants; (4.18) becomes:

$$E = E^0 + \frac{0 \cdot 059}{n} \log \frac{a_A^a \cdot a_B^b \cdot a_C^c}{a_K^k \cdot a_L^l \cdot a_M^m}. \tag{4.20}$$

4.6. Electrode potentials and equilibrium constants. According to (4.17), the multiplication of the potentials in table 4.2 by $-23 \cdot 06$ yields the standard free energy change per electron of the corresponding reaction in kcals. As discussed in §4.5, the electron is an abbreviation for the change $[\frac{1}{2}H_2(g) - H^+(aq)]$ at unit activity, and these free energies therefore refer to a series of stoicheiometrically equivalent reactions. Hence, subtraction of two potentials followed by multiplication by $-23 \cdot 06$ yields the standard free energy change of the subtracted reactions, reactants and products being multiplied by coefficients giving equivalence to one electron. These standard free energies may be readily converted into equilibrium constants:

$$IO_3^- + 6H^+ + 5e \to \tfrac{1}{2}I_2 + 3H_2O, \quad E^0 = 1 \cdot 19 \text{ V}, \tag{4.12}$$

$$\tfrac{1}{2}I_2 + e \to I^-, \quad E^0 = 0 \cdot 54 \text{ V}, \tag{4.13}$$

$$\therefore \tfrac{1}{5}IO_3^- + I^- + \tfrac{6}{5}H^+ \to \tfrac{6}{10}I_2 + \tfrac{3}{5}H_2O, \quad \Delta G^0 = -14 \cdot 98 \text{ kcals}, \tag{4.14}$$

$$IO_3^- + 5I^- + 6H^+ \to 3I_2 + 3H_2O, \quad \Delta G^0 = -74 \cdot 90 \text{ kcals}, \tag{4.15}$$

$$\therefore K = 9 \times 10^{54}. \tag{4.21}$$

The large value of this constant implies that the reaction is virtually complete, a fact that is used extensively in volumetric analysis for the standardization of thiosulphate solutions.

A slightly different approach may be demonstrated with the reaction:

$$Sn^{2+} + 2Fe^{3+} \rightarrow Sn^{4+} + 2Fe^{2+}. \qquad [4.16]$$

If we imagine that the reaction is carried out in the cell:

$$^-Pt|Sn^{2+}, Sn^{4+}|salt\ bridge|Fe^{2+}, Fe^{3+}|Pt^+,$$

then relative to the standard hydrogen electrode, the potential of the wire in the tin solution is given by

$$E_{Sn} = 0 \cdot 15 + \frac{0 \cdot 059}{2} \log \frac{a_{Sn^{4+}}}{a_{Sn^{2+}}}, \qquad (4.22)$$

while that in the iron solution is:

$$E_{Fe} = 0 \cdot 77 + 0 \cdot 059 \log \frac{a_{Fe^{3+}}}{a_{Fe^{2+}}}. \qquad (4.23)$$

If the cell is reversible, current stops flowing at equilibrium when the e.m.f. is zero. Here:

$$E_{Sn} = E_{Fe} \qquad (4.24)$$

and

$$0 \cdot 059 \log \frac{(a_{Sn^{4+}})^{\frac{1}{2}} \cdot a_{Fe^{2+}}}{(a_{Sn^{2+}})^{\frac{1}{2}} \cdot a_{Fe^{3+}}} = 0 \cdot 62, \qquad (4.25)$$

$$\therefore K = \frac{a_{Sn^{4+}} \cdot (a_{Fe^{2+}})^2}{a_{Sn^{2+}} \cdot (a_{Fe^{3+}})^2} = 1 \times 10^{21}. \qquad (4.26)$$

Again the equilibrium lies almost totally to the right, and the reaction is used to reduce iron in ores to the ferrous state, prior to its estimation with dichromate.

It should be clear from the preceding calculations that, in the one electron case, a difference of only $0 \cdot 059$ V in two potentials implies an equilibrium constant of ten in the reaction obtained by combining them while, for more than one electron, the equilibrium constant will be even larger.

Table 4.2 contains in a concise form all the material necessary for the calculation of 528 such equilibrium constants.

Within the limits of accuracy of ordinary analytical techniques, the combined reaction may be regarded as complete if the individual potentials differ by more than 0·4 V.

4.7. Some limitations of standard electrode potentials.

§4.6 shows that electrode reactions can be arranged in order of descending E^0 such that, under standard conditions, the oxidized state of any couple going down to the reduced state is thermodynamically capable of oxidizing the reduced state of any couple beneath it to the oxidized state. This sentence includes two very important qualifications, one of which is implicit in the phrase 'under standard conditions'.

The equilibrium constants derived in the preceding section are functions of activities, but it is the equilibrium position in terms of concentrations which interests the inorganic chemist. Only at very low ionic strengths, where the activity coefficients are unity, do the concentration and activity equilibrium constants coincide. Fortunately, the activity coefficients of electrolytes rarely differ by more than a factor of ten in normal test tube media, and the activity equilibrium constant is usually a good approximation (within a factor of 100) to its equilibrium counterpart. Under these conditions then, the standard electrode potential is an excellent guide to the oxidizing or reducing capacity of a particular system. The reaction:

$$IrBr_6^{2-} + Br^- \rightarrow IrBr_6^{3-} + \tfrac{1}{2}Br_2 \qquad [4.17]$$

is an interesting exception. The standard electrode potentials of the couples $IrBr_6^{2-}/IrBr_6^{3-}$ and $\tfrac{1}{2}Br_2(aq)/Br^-$ are 0·99 V and 1·09 V, and they suggest that the equilibrium lies well to the left. This is the correct conclusion at low ionic strengths, but if a considerable quantity of potassium nitrate is added, liberation of bromine and reduction of hexabromoiridate (IV) occurs. Denoting the concentration and activity equilibrium constants by K_c and K_a, and assuming that the activity coefficient of bromine is unity,

$$K_c = K_a \cdot \frac{\gamma_{IrBr_6^{2-}} \cdot \gamma_{Br^-}}{\gamma_{IrBr_6^{3-}}}. \qquad (4.27)$$

At very low values of the ionic strength, μ, the Debye–Hückel limiting law holds and

$$\log \gamma = -0\cdot51z_i^2\sqrt{\mu}, \tag{4.28}$$

where z_i is the charge on an ion.

$$\therefore \log K_c = \log K_a + 2\cdot04\sqrt{\mu}. \tag{4.29}$$

Thus the concentration equilibrium constant increases with the ionic strength at low values of the latter. It is evident from the experimental results that this trend continues beyond those ionic strengths at which the Debye–Hückel limiting law holds.

Like all thermodynamic data, electrode potentials are subject to the limitations imposed by kinetics. A reaction may be thermodynamically possible but large activation energies may prevent it from proceeding at an observable rate. Thus when magnesium ribbon is dipped in cold water, a table of electrode potentials shows that evolution of hydrogen is thermodynamically feasible, but no reaction is observed because a thin film of oxide on the magnesium imposes a large activation energy barrier to the contact of the metal and the liquid. If the coherence of the oxide layer is destroyed by amalgamation, the activation energy is lowered and a brisk reaction occurs.

Reactions which appear to involve only transfer of electrons are usually fast; thus the oxidation of ferrous by ceric ions is apparently instantaneous, but those which obviously involve the making and breaking of covalent bonds vary enormously in speed. The ClO_4^-/Cl^- potential exceeds that of the $\frac{1}{2}Cr_2O_7^{2-}/Cr^{3+}$ couple, but although ferrous ions are stable for long periods in de-aerated perchloric acid, they are instantly oxidized by dichromate.

The oxidation of manganese (II) by persulphate in sulphuric acid media is extremely slow, but if a little silver nitrate is added and the solution warmed, persulphate oxidizes the silver to Ag^{3+} which acts as an oxidizing intermediate for the rapid production of permanganate. This catalytic effect is used in the volumetric determination of manganese. Phosphoric acid is usually added to complex with the manganese (III) intermediate. This prevents the precipitation of manganese dioxide and the permanganate may be estimated by the oxalate method when excess persulphate has been destroyed by heating.

4.8. Disproportionation. An aquated molecule or ion can sometimes decompose by disproportionation, a process of simultaneous oxidation and reduction which frequently occurs without the intervention of an extraneous species. This reaction may be represented by the equation:

$$(y+z)M^x \rightarrow yM^{x+z} + zM^{x-v}. \tag{4.18}$$

If $\qquad E^0[M^{x+z}/M^x] = k \quad$ and $\quad E^0[M^x/M^{x-v}] = l,$

Then $\qquad\qquad \Delta G^0 = -zyFl + yzFk \tag{4.30}$

and will be negative if $l > k$.

Thus, if an element forms a lower, an intermediate and a higher oxidation state in aqueous solution, and the potential of the couple involving the higher and intermediate states is less than that of the couple containing the intermediate and lower states, then the intermediate species is unstable to disproportionation into those containing the element in the lower and higher oxidation states.

Thus $E^0[Cu^{2+}/Cu^+] = 0.15$ V while $E^0[Cu^+/Cu] = 0.52$ V, and the dissolution of cuprous sulphate in water is followed by the precipitation of metallic copper and the formation of cupric ions. On the other hand, $E^0[Fe^{3+}/Fe^{2+}] = 0.77$ V while $E^0[Fe^{2+}/Fe] = -0.44$ V, and the addition of iron to a solution of excess ferric chloride results in total conversion of the solid metal to the ferrous state. This process, followed by a titration with standard dichromate, formed the basis of an old method for the determination of elemental iron.

Many disproportionation equilibria, especially those involving oxyanions, are highly sensitive to pH. The standard potential of the couple:

$$BrO_3^- + 6H^+ + 5e \rightarrow \tfrac{1}{2}Br_2 + 3H_2O \tag{4.19}$$

is 1.52 V, but at a pH of 10, when the remaining species are at unit activity, the potential E is given by

$$E = 1.52 + \frac{0.059}{5} \log a_{H^+}^6 \tag{4.31}$$

$$= 0.81 \text{ V}. \tag{4.32}$$

Since $E^0[\frac{1}{2}Br_2/Br^-] = 1\cdot08$ V, in acid solution bromine is stable with respect to disproportionation into bromate and bromide, but unstable at a pH of 10. At an important stage in the manufacture of bromine from sea water, the gas is absorbed in sodium carbonate solution when it forms bromide and bromate. Subsequent acidification reliberates it in concentrated form:

$$BrO_3^- + 5Br^- + 6H^+ \to 3Br_2 + 3H_2O. \qquad [4.20]$$

This equilibrium is an excellent example of the variations in the stabilities of oxidation states with pH. Further instances are discussed in the next section. Volumetric analysis details are from Kolthoff and Belcher (1957).

4.9. The stabilities of oxidation states.

Suppose that a metal forms a complex in two different oxidation states, b and c, with the same ligand. Then a solution of both complexes, each at unit activity, will contain a usually very small but finite concentration of the ions M^{b+} and M^{c+}. If a platinum wire is dipped into the solution, then the system may be regarded either as a standard electrode with respect to the complexes, or as an M^{c+}/M^{b+} electrode whose potential differs markedly from the standard value. This difference can be determined if the instability constants for the dissociation of the complexes into their respective aquo-ions are known, for from these constants and the ligand concentration prevailing in the solution, the equilibrium concentrations of the ions M^{c+} and M^{b+} can be evaluated. Then:

$$E^0_{comp} = E_{aq} = E^0_{aq} + \frac{0\cdot059}{c-b} \log \frac{[M^{c+}]}{[M^{b+}]}, \qquad (4.33)$$

where $c > b$ and, as in the remainder of this chapter, molal concentrations have been substituted for activities. Since E^0_{comp} is a satisfactory measure of the relative stability of the complexes, the variation in these stabilities as the ligand changes may be discussed in terms of the potential of the aquo-couple and the overall instability constants of the complexes. The oxidation state of a metal in a complex is usually defined as the charge left on the metal when the ligands are removed in their closed shell configurations. In this form they are usually stable in solution, so their concentrations may be measured and controlled. Consequently the measurement of the dissociation constant involving a complex with the metal in an oxidation state c and the aquo ion, M^{c+}, is made much easier. This is one of the most attractive features of the definition of oxidation state—it arbitrarily systematizes coordination chemistry in a way that makes the quantitative data necessary for the discussion of the variation of stabilities of oxidation states much easier to obtain. Since instability constants may vary by as much as 10^{50}, it is clear that, despite the logarithmic dependence of (4.33), the behaviour of an oxidation state may change very markedly in the presence of different ligands whose addition to the solution causes the precipitation of insoluble compounds or the formation of stable complexes.

A glance at table 4.2 shows that many electrode potentials are dependent on the hydrogen ion concentration which may vary in water by a factor of about 10^{14}. Thus a change in pH, even where it does not alter the nature of the complex, may also cause a marked alteration in the behaviour of particular oxidation states. The recognition of these phenomena is of the greatest importance in the interpretation of inorganic chemistry. Some illustrative examples now follow.

The iron (III)–iodide equilibrium. The standard potentials of the couples Fe^{3+}/Fe^{2+} and $\frac{1}{2}I_2/I^-$ are 0·77 V and 0·54 V respectively. In neutral or acid solution, in accord with these values, ferric ion will oxidize iodide to iodine. On the other hand, if air is excluded to prevent side reactions and a ferrous solution is made alkaline, the hydroxide precipitated is readily oxidized to ferric hydroxide by iodine.

Consider a mixture of ferrous and ferric hydroxides in contact with a molal solution of hydroxide ions. The solubility products are about 10^{-15} and 10^{-38} respectively, and in the given solution, these minute figures represent the equilibrium concentrations of ferrous and ferric ions.

$$\therefore E = 0·77 + 0·059 \log 10^{-38}/10^{-15} \tag{4.34}$$

$$= -0·59 \text{ V.} \tag{4.35}$$

This figure is equal to the standard potential of the couple:

$$Fe(OH)_3 + e = Fe(OH)_2 + OH^-. \tag{4.21}$$

It lies well below the iodine potential, so iodine is perfectly capable of oxidizing ferrous hydroxide in alkaline solution. The figures may be qualitatively interpreted by saying that the solubility of ferric hydroxide is so small that the equilibrium position for the reaction:

$$\text{iron (III)} + I^- \to \text{iron (II)} + \tfrac{1}{2}I_2 \qquad [4.22]$$

lies well over to the left. A similar effect may be produced by the addition of fluoride rather than hydroxide ions. These yield a stable complex FeF_6^{3-} and prevent the oxidation of iodide, but HF is so weak an acid that, on addition of hydrogen ions, the complex is broken down and the equilibrium position moves well to the right. The addition of fluoride ions enables copper to be determined by the direct iodometric method in ores contaminated with ferric iron.

If a molal solution of ferrous and ferric ions is made molal in cyanide, then stable complexes $Fe(CN)_6^{4-}$ and $Fe(CN)_6^{3-}$ are eventually formed in both oxidation states. The ratio of their instability constants is 10^7,[1] so in this solution:

$$E^0_{comp} = E_{aq} = 0 \cdot 77 + 0 \cdot 059 \log 10^{-7}, \qquad (4.36)$$

$$= 0 \cdot 36 \text{ V}. \qquad (4.37)$$

Thus in neutral conditions, iodine will oxidize ferrocyanide to ferricyanide, and if a dilute buffered solution of the complex is used with a large excess of the halogen, volumetric determination of ferrocyanide is possible by back titration with standard thiosulphate.

On the other hand, because ferrocyanic acid is weak, the addition of hydrogen ions to an equimolar mixture of the cyanides lowers the concentration of ferrocyanide by a factor of about 10^{-7}. The potential of the system then becomes roughly $0 \cdot 8$ V, and ferricyanide now oxidizes iodide to iodine. In acid solution therefore, the reaction may be used for the volumetric determination of ferricyanide. If acid conditions are inconvenient, addition of zinc and potassium ions causes precipitation of an insoluble ferrocyanide, $K_2Zn_3[Fe(CN)_6]_2$ which also shifts the equilibrium in [4.22] to the right.

The oxidation states of copper. On page 80, the tendency of cuprous compounds to disproportionate was pointed out. This,

---

[1] Although here the ratio is introduced first for didactic purposes, the figure is in fact obtained from the electrically measured standard potentials of the cyano- and aquo-complex systems.

allied with the fact that $E^0[Cu^{2+}/Cu^+]$ is only $0 \cdot 15$ V, suggests that cupric ions should be incapable of oxidizing iodide to iodine when Cu^+ is the other product. In the presence of excess iodide however, cuprous ions form insoluble CuI which has a solubility product of 10^{-12}. Thus,

$$E^0[Cu^{2+}/CuI] = 0 \cdot 15 + 0 \cdot 059 \log 10^{12}, \qquad (4.38)$$

$$= 0 \cdot 86 \qquad (4.39)$$

and the process

$$Cu^{2+} + 2I^- \rightarrow CuI + \tfrac{1}{2}I_2 \qquad [4.23]$$

occurs readily. This reaction interferes with the iodometric estimation of potassium iodate when cupric ions are present. If excess sodium citrate is added, copper(II) is stabilized by the formation of a citrate complex, and subsequent addition of potassium iodide, acidified with acetic acid, yields iodine equivalent only to the iodate. When this has been titrated with thiosulphate, addition of a strong acid breaks up the citrate complex with formation of the weak citric acid, and precipitation of cuprous iodide occurs. Titration of the fresh iodine liberated by this reaction then yields the copper content of the original iodate solution.

In concentrated hydriodic acid, copper(I) is even further stabilized by dissolution of cuprous iodide to give the complex CuI_2^-. Although $E^0[Cu^+/Cu]$ is $0 \cdot 52$ V, copper will dissolve in concentrated hydriodic acid with evolution of hydrogen to form the complex, because in this medium,

$$E[CuI_2^-/Cu] < E^0[CuI/Cu] = -0 \cdot 19 \text{ V}. \qquad (4.40)$$

Undoubtedly too, the reaction is made more favourable by an increase in the potential of the hydrogen electrode caused by the large concentration and activity coefficient of hydrogen ions in strong hydriodic acid. Dilution of the final solution precipitates cuprous iodide. Similar behaviour is observed for silver and mercury. The formula of the silver complex is not well established, but for mercury it is HgI_4^{2-}. The standard potentials of the couples Ag^+/Ag and Hg^{2+}/Hg are $0 \cdot 80$ V and $0 \cdot 85$ V.

The oxidation states of plutonium. The complicated nature of the aqueous chemistry of plutonium is concisely expressed by the potential diagram in fig. 4.2.

This implies, correctly, that in molal acid solution, plutonium-(IV) is just stable to disproportionation into plutonium(III) and plutonium(VI):

$$3Pu^{4+} + 2H_2O \rightarrow 2Pu^{3+} + PuO_2^{2+} + 4H^+ \qquad [4.24]$$

while plutonium(V) is just unstable to plutonium(IV) and plutonium(VI):

$$2PuO_2^+ + 4H^+ \rightarrow Pu^{4+} + PuO_2^{2+} + 2H_2O. \qquad [4.25]$$

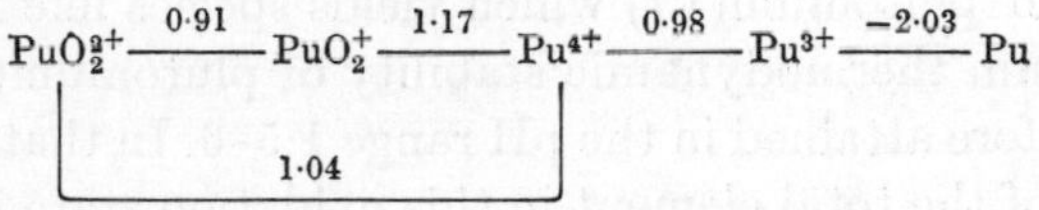

$$PuO_2^{2+} \xrightarrow{0.91} PuO_2^+ \xrightarrow{1.17} Pu^{4+} \xrightarrow{0.98} Pu^{3+} \xrightarrow{-2.03} Pu$$

$$\underset{1.04}{\underline{\qquad\qquad\qquad}}$$

Fig. 4.2. The oxidation states and electrode potentials for plutonium.

Thus, when a little plutonium(IV) is dissolved in 0·5M HCl at 25°, the equilibrium solution contains appreciable proportions of all four oxidation states; 27·2 per cent Pu^{III}, 58·4 per cent Pu^{IV}, 13·6 per cent Pu^{VI} and 0·75 per cent Pu^V. This situation may be very much altered by complex ion formation, and the quadruply charged ions of the lanthanides and actinides are especially interesting because they demonstrate the capacity of the common anions like halide, sulphate and nitrate to form stable complexes. Of these anions, perchlorate is much the weakest complexing agent, while sulphate is one of the best. Thus in M H_2SO_4, the disproportionation of plutonium(IV) is negligible.

As (4.7) indicates, the molal entropy of Pu^{4+}(aq) is large and negative due to its high charge and fairly small size. This imparts a standard entropy change of over 100 cal/deg.mole to [4.24], and the equilibrium constant increases by roughly a thousand times on raising the temperature a mere 45° (cf. page 10). The hydrogen ion dependence of [4.24] suggests that the disproportionation should be favoured by addition of bases. This is observed until the pH becomes about two, when precipitation of the very insoluble plutonium(IV) hydroxide results in the net reaction:

$$3Pu(OH)_4 + 8H^+ \rightarrow 2Pu^{3+} + PuO_2^{2+} + 10H_2O. \qquad [4.26]$$

Thus at higher values of the pH, the disproportionation reaction is negligible.

[**4.25**] correctly implies that the equilibrium concentration of plutonium(V) may be increased by using solutions containing non-complexing anions like perchlorate to destabilize plutonium-(IV), and by raising the pH. When the latter reaches about 1·5, the polymerization of plutonium(IV) and its precipitation as the hydroxide renders the reaction acid-independent:

$$2PuO_2^+ + 2H_2O \rightarrow Pu(OH)_4 + PuO_2^{2+}, \qquad [\textbf{4.27}]$$

but at a pH higher than 6, disproportionation is enhanced by the hydrolysis of plutonium(VI) which yields species like $Pu(OH)O_2^+$. The maximum thermodynamic stability of plutonium(V) in solution is therefore attained in the pH range 1·5–6. In that region, the proportion of the total element in this oxidation state is increased at low values of the total plutonium concentration, because the equilibrium constant of [**4.27**] depends on the inverse of the *square* of the activity of plutonium(V).

By using these facts, it is possible to obtain stable solutions containing over 90 per cent of their plutonium in the pentapositive state.

More extensive quantitative data on the complexing effects of the common anions are available for cerium compounds.

In M perchloric, nitric and sulphuric acids, the potentials of solutions containing small but equal concentrations of cerium(III) and cerium(IV) are 1·75, 1·61 and 1·43 V respectively, corresponding with increasing complexing of the +4 state. If a few very small crystals of manganese perchlorate are added to excess cerium(IV) in perchloric or nitric acid, the purple colour of permanganate, which may be distinguished from Mn^{III} by its visible absorption spectrum, is readily discernible on standing. On the other hand, permanganate brings about slow but complete oxidation of cerium-(III) in dilute sulphuric acid. The reaction is accelerated by the addition of silver ions (cf. the persulphate oxidation on page 79).

4.10. Solvent decomposition. The range of chemical species found in aqueous solution is naturally restricted to those which do not rapidly oxidize or reduce the solvent. The first process almost invariably yields oxygen and, in acid solution, the relevant potential is:

$$O_2 + 4H^+ + 4e \rightarrow 2H_2O, \quad E^0 = 1·23 \text{ V}. \qquad [\textbf{4.28}]$$

With unit pressure of oxygen, the variation of this potential with the hydrogen ion concentration reduces to:

$$E = 1\cdot23 - 0\cdot059 \text{ pH}, \tag{4.41}$$

so in alkaline solution at a pH of 14, $E = 0\cdot40$ V.

Solutions normally exist in contact with atmospheric oxygen, and air-sensitive ions such as V^{2+} and Cr^{2+} are unstable with respect to the reverse of the water oxidation reaction. This gives [4.28] additional importance. Indeed, if thermodynamics alone were important and no precautions were taken to exclude dissolved gases, at unit activity of hydrogen ions, the oxidized states of all couples with potentials greater than 1.23 V would oxidize water, and the reduced states of all couples with potentials below this value would be oxidized by dissolved oxygen. Under these circumstances, ions such as permanganate ($E^0[MnO_4^-/Mn^{2+}] = 1\cdot51$ V) and cerium(IV); ($E^0[Ce^{IV}/Ce^{III}] = 1\cdot43$ V in dilute sulphuric acid) would not exist in aqueous solution, but species such as Fe^{3+} ($E^0[Fe^{3+}/Fe^{2+}] = 0\cdot77$ V) would be stable because the production of only a minute quantity of the dipositive state would be necessary to give a potential of $1\cdot23$ V. Substantial quantities of ferrous ions would not be capable of a sustained existence owing to their sensitivity to dissolved oxygen. In practice, kinetic factors which are presumably connected with the strengths of the bonds in molecular oxygen and water, make this a hypothetical situation. Thus Latimer (1952) observed that the potential of a couple must exceed that of the oxygen electrode by roughly at least 0·6 V before its oxidized state decomposes water rapidly. This quantity is sometimes called the oxygen overvoltage by analogy with the excess e.m.f. over the reversible potential that is required to liberate the gas at most metallic anodes. The figure should not be taken too seriously. It implies a relation between the kinetics and the thermodynamics of the oxidation of water that is improperly understood.

At the other end of the scale, reduction of water proceeds via the reaction:

$$H^+(aq) + e \rightarrow \tfrac{1}{2}H_2(g), \quad E^0 = 0 \tag{4.29}$$

and with unit fugacity of hydrogen, the potential of this electrode is given by:

$$E = -0\cdot059 \text{ pH}. \tag{4.42}$$

Potentials about 0·6 V less than this are usually required to produce hydrogen, and by analogy with the barrier to the reduction of water at electrodes, this quantity is again called the overvoltage. The hydrogen overvoltage probably arises from the difficulty of producing hydrogen atoms from aqueous hydrogen ions in the rate-determining step, an energy barrier which is reflected in the large negative value of the $H^+(aq)/H(g)$ potential ($-2\cdot1$ V). This activation energy is substantially lowered by those metals which form surface hydrides with atomic hydrogen. In the presence of platinized platinum the overvoltage is almost nil, and, as we have seen, when this material is used as an electrode, the hydrogen couple becomes reversible. Furthermore, the addition of finely divided platinum to aqueous solutions of reducing agents like V^{2+} and Cr^{2+} will frequently catalyse the reduction of the solvent if the process is thermodynamically feasible.

Thus just as in the oxidation of water, kinetic effects increase the potential required for rapid reaction by about 0·6 V, so in the case of reduction it lowers it by about the same amount. It follows that, although the *thermodynamic* stability of oxidizing or reducing agents in aqueous solution is such that the potentials of their couples must lie between $-0\cdot059$ pH and $(1\cdot23-0\cdot059$ pH$)$ volts, the range is roughly doubled by kinetic factors. This fortunate occurrence greatly extends the interest and variety of chemistry in aqueous solution.

The lowering of both the extreme potentials for thermodynamic stability in water as the pH is raised, suggests that low oxidation states might be stabilized at the expense of higher ones in alkaline media. In fact the converse is generally true. High oxidation states in acid solution often exist as aquo-ions, oxy-cations or oxy-anions. When solutions of the first two are made alkaline, oxides or hydroxides are usually precipitated, and their solubilities commonly diminish very markedly with increasing oxidation number of the metal.

In the case of oxy-anions, which may be regarded as hydroxide complexes from which water has been removed, the electrode potential for reduction to a lower oxidation state is nearly always highly sensitive to pH:

$$MnO_4^- + 8H^+ + 5e \rightarrow Mn^{2+} + 4H_2O. \qquad [4.30]$$

Consequently, in either of these two situations, the lowering of the potential of a couple containing two oxidation states when the solution is made alkaline is usually even greater than that in the potentials appropriate to the oxidation or reduction of water. Under these circumstances, lower oxidation states are more likely to reduce the solvent, to disproportionate, or to be oxidized by dissolved air. For example, acidified de-oxygenated solutions of titanium(III) sulphate are stable indefinitely, but when the hydroxide is precipitated by addition of alkali, rapid evolution of hydrogen occurs with formation of a hydrated oxide of titanium-(IV). Chlorine water contains appreciable quantities of the dissolved gas, but if the halogen is passed into boiling alkali, rapid disproportionation to chloride and chlorate takes place. The standard potential of the couple Co^{3+}/Co^{2+} is 1·95 V, well above the figure of 1·23 V required for the oxidation of water in molal acid, and in this solution, cobaltic ions rapidly decompose the solvent. In molal alkali, from (4.41), the oxidation of water requires a potential of only 0·40 V, but the hydroxides $Co(OH)_2$ and $Co(OH)_3$ have solubility products of about 10^{-15} and 10^{-50} respectively. A calculation analogous to that for iron on page 82 shows that under these conditions, the cobalt(III)–cobalt(II) potential is as low as −0·12 V and dissolved oxygen now oxidizes cobalt(II) hydroxide to the tripositive state. Again, ferrates are made by the hypochlorite oxidation of ferric hydroxide in strong alkali, but when dropped into acid they instantly oxidize the solvent.

4.11. Non-aqueous solvents. Liquid ammonia is the only non-aqueous solvent in which the thermodynamics of chemical reactions have been appreciably studied. A brief look at this work is very instructive, for it shows clearly that chemical behaviour in one medium may differ substantially from that in another.

In water, the scale of oxidation potentials was defined by assigning zero e.m.f. to the standard hydrogen electrode, i.e. by defining the electron symbol in table 4.2 as a form of shorthand for $[\frac{1}{2}H_2(f = 1) - H^+(aq., a = 1)]$. In liquid ammonia, the corresponding scale is referred to the electrode made up of hydrogen gas at unit fugacity in equilibrium with a standard solution of ammonium ions in liquid ammonia. Some standard electrode potentials at 25° in this medium are recorded in table 4.3. Values are from Jolly (1956) and have mostly been obtained by correcting

T ABLE 4.3   *Some standard electrode potentials in liquid ammonia at 25°*

Electrode	E^0 (Volts)
$Li + e \rightarrow Li$	$-2{\cdot}34$
$Sr^{2+} + 2e \rightarrow Sr$	$-2{\cdot}3$
$Ba^{2+} + 2e \rightarrow Ba$	$-2{\cdot}2$
$Ca^{2+} + 2e \rightarrow Ca$	$-2{\cdot}17$
$Cs^+ + e \rightarrow Cs$	$-2{\cdot}08$
$Rb^+ + e \rightarrow Rb$	$-2{\cdot}06$
$K^+ + e \rightarrow K$	$-2{\cdot}04$
$e \rightarrow e^-(am)$	$-1{\cdot}95$
$Na^+ + e \rightarrow Na$	$-1{\cdot}89$
$Mg^{2+} + 2e \rightarrow Mg$	$-1{\cdot}74$
$Zn^{2+} + 2e \rightarrow Zn$	$-0{\cdot}54$
$NH_4^+ + e \rightarrow NH_3 + \frac{1}{2}H_2$	0
$\frac{1}{2}N_2 + 3NH_4^+ + 3e \rightarrow 4NH_3$	$0{\cdot}04$
$Cu^+ + e \rightarrow Cu$	$0{\cdot}36$
$Cu^{2+} + 2e \rightarrow Cu$	$0{\cdot}40$
$Hg^{2+} + 2e \rightarrow Hg$	$0{\cdot}67$
$Ag^+ + e \rightarrow Ag$	$0{\cdot}76$
$\frac{1}{2}I_2 + e \rightarrow I^-$	$1{\cdot}26$
$\frac{1}{2}Br_2 + e \rightarrow Br^-$	$1{\cdot}73$
$\frac{1}{2}Cl_2 + e \rightarrow Cl^-$	$1{\cdot}91$
$\frac{1}{2}F_2 + e \rightarrow F^-$	$3{\cdot}50$

measurements made at $-33°$, the boiling point of liquid ammonia. In most cases, qualitative conclusions deduced from table 4.3 are also valid at $-33°$. The electron symbol is now an abbreviation for $[NH_3(l) + \frac{1}{2}H_2(g, f = 1) - NH_4^+(am, a = 1)]$.

In fig. 4.3, the thermodynamic factors which contribute to the order of potentials of the couples M^{n+}/M and $\frac{1}{2}X_2/X^-$ in any solvent are shown. Small terms in RT have been omitted.

ΔG_s^0 represents the standard free energy of solvation and the other symbols have their usual meanings. In (a), the overall free energy change is equal to $n(FE^0 + \text{const.})$ while in (b) it is $-(FE^0 + \text{const.})$. The electron symbol here represents an electron in the gas phase, and for each solvent the constant is a function of the reference electrode of zero e.m.f.

In water, the relatively low sublimation energies and ionization potentials of the alkali metals and alkaline earths cause low E^0

values. These energy terms are independent of the solvent, so that it is not surprising to find that the same metals are very powerful reducing agents in liquid ammonia. For the first step in (a), the order of values of $\Delta G^0/n$ is

$$\text{Cs} < \text{Rb} < \text{K} < \text{Na} < \text{Li} < \text{Ba} < \text{Sr} < \text{Ca} < \text{Mg}.$$

At 25°, the dielectric constant of ammonia is 16·9 compared with 78·5 for water. If the Born expression for free energies of solvation

(a) $\text{M(s)} \xrightarrow[-T\Delta S_I^0[\text{M}^{n+}(\text{g})]]{L_V + \overset{n}{\underset{0}{\sum}} I_n} \text{M}^{n+}(\text{g}) + ne^-(\text{g}) \xrightarrow{\Delta G_s^0[\text{M}^{n+}]} \text{M}^{n+}(\text{aq}) + ne^-(\text{g})$

(b) $\frac{1}{2}\text{X}_2(\text{ref. state}) + e^-(\text{g}) \xrightarrow[-T\Delta S_I^0[\text{X}^-(\text{g})]]{\Delta H_I^0[\text{X}^-(\text{g})]} \text{X}^-(\text{g}) \xrightarrow{\Delta G_s^0[\text{X}^-]} \text{X}^-(\text{aq})$

Fig. 4.3. A breakdown of the standard free energies of
(a) $\text{M(s)} \rightarrow \text{M}^{n+}(\text{aq}) + ne^-(\text{g})$ and (b) $\frac{1}{2}\text{X}_2 + e^-(\text{g}) \rightarrow \text{X}^-(\text{aq})$.

were correct (see page 103), the smaller differences between them in the medium of lower dielectric constant should cause the order of potentials to approach more nearly that of $\Delta G^0/n$. That the electrode potentials of the alkaline earths diminish, relative to those of the alkali metals, in moving from water to liquid ammonia shows the weakness of the continuous dielectric approach and attests to the more subtle factors which must influence the strength of the solvent interactions with specific ions. The relatively higher position of the alkaline earths in the liquid ammonia table is not surprising in view of the formation of stable solid ammines, $\text{M(NH}_3)_6\text{X}_2$, by these elements.

For the halogens, the standard free energies and enthalpies of the first stage in fig. 4.3 (b) are surprisingly similar (see table 2.10), so the order of potentials in the two media is determined by the free energies of solvation. As expected, these increase with the ionic radius, and the oxidizing power of the halogens diminishes from fluorine to iodine. The increase in the electrode potential in moving from water to liquid ammonia is 0·1 V larger for iodine than for the remaining halogens. This suggests that relative to water, the iodide ion coordinates more strongly with liquid ammonia than the other anions, an observation that is in accord with the noticeable solubility of iodides in the latter solvent.

The rough similarity of the 'electrochemical series' in the two media does not extend to reactions involving a more subtle interplay of solvation energies and ionization potentials. The potential diagram for mercury in an acid solution of liquid ammonia is shown in fig. 4.4, and it is clear that the mercurous ion is highly unstable with respect to disproportionation. The ready solubility of mercuric chloride and bromide in liquid ammonia compared with solubility products of about 10^{-20} in water, and an overall in-

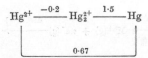

Fig. 4.4. Potential diagram for mercury in liquid ammonia at 25°.

stability constant of 10^{-20} for the aquated $Hg(NH_3)_4^{2+}$ complex reflect the strong interaction of mercuric ions with ammonia which produces this interesting result. On the other hand, cuprous ions are stable to disproportionation in liquid ammonia. Similar behaviour is observed in acetonitrile where silver ions will oxidize copper to the cuprous rather than the cupric state.

In an acid solution of liquid ammonia, the reducing agents in couples with negative potentials are thermodynamically capable of reducing the solvent:

$$NH_4^+ + e \rightarrow \tfrac{1}{2}H_2 + NH_3, \quad E^0 = 0 \tag{4.31}$$

As in water, a surprising number of such agents exist, and it would seem that the ability of liquid ammonia to stabilize compounds such as $K_4Ni(CN)_4$, KPH_2 and $NaGeH_3$ is mainly due to a kinetic barrier to reduction of the solvent. The overvoltage for rapid completion of this process is usually placed as high as $1 \cdot 2$ V, although naturally the figure is prone to vary in specific cases. This still does not make liquid ammonia an impressive reduction-resistant medium in acid solution, but the equilibrium constant for the ionization of the solvent:

$$2NH_3 \rightarrow NH_4^+ + NH_2^- \tag{4.32}$$

is as low as 10^{-27} at 25°. Thus in 'neutral' ammonia the ammonium ion concentration is $10^{-13 \cdot 5}$, and the potential associated with [4.31] falls to $-0 \cdot 80$ V compared with only $-0 \cdot 41$ V in water.

This, combined with the high overvoltage, is sufficient to make liquid ammonia an excellent solvent for reducing reactions. Its excellence is accentuated by the availability of an ideal reducing agent. The ammoniated electron, whose standard potential of $-1\cdot95$ V refers to the reverse of the reaction:

$$e^-(\text{am}) + \text{NH}_4^+(\text{am}, a = 1) \to \text{NH}_3(\text{l}) + \tfrac{1}{2}\text{H}_2(\text{g}, f = 1) \qquad [4.33]$$

is a powerful reducing species which forms a homogeneous phase with the solvent, and in cold neutral solution reduces it only slowly. The generation of ammoniated electrons may be carried out by adding to the solvent a reducing agent whose couple has a standard potential of less or just greater than $-1\cdot95$ V. Thus for the alkali metals and the alkaline earths, the reaction:

$$\text{M}(\text{s}) \to \text{M}^{n+}(\text{am}) + n e^-(\text{am}) \qquad [4.34]$$

has a standard free energy which is negative or nearly so, and the dissolution of small quantities of these elements in liquid ammonia yields the well known blue solutions of the ammoniated electron. The addition of ammonium ions raises the potential required for the reduction of the solvent to about zero volts, so the hydrogen overvoltage is exceeded and the solutions are decolourized via [4.33]. As expected, the measured standard enthalpy of this process is independent of the original metal dissolved. Tetra-alkyl-ammonium ions have no effect. Indeed, thanks to the hydrogen overvoltage, ammoniated electrons can be generated by cathodic reduction of their salts in liquid ammonia.

It is interesting that recent pulse radiolysis studies have proved that the hydrated electron has a transient existence in water. Kinetic work suggests that $\Delta G^0 = 8\cdot4$ kcal/mole for the reaction:

$$\text{H}_2\text{O}(\text{l}) + e^-(\text{aq}) \to \text{H}(\text{aq}) + \text{OH}^-(\text{aq}). \qquad [4.35]$$

Combining an estimated free energy of hydration of the hydrogen atom ($+4\cdot5$ kcal/g-atom) with the free energies of dissociation of hydrogen,

$$\tfrac{1}{2}\text{H}_2(\text{g}) \to \text{H}(\text{g}), \quad \Delta G^0 = 48\cdot6 \text{ kcal} \qquad [4.36]$$

and water,

$$\text{H}_2\text{O}(\text{l}) \to \text{H}^+(\text{aq}) + \text{OH}^-(\text{aq}), \quad \Delta G^0 = 19\cdot1 \text{ kcal} \qquad [4.37]$$

we obtain 63·8 kcal/mole for the standard free energy of the reaction,

$$\tfrac{1}{2}H_2(g) \rightarrow H^+(aq) + e^-(aq) \tag{4.38}$$

at 25°. This gives $-2\cdot77$ V for the standard potential of the hydrated electron. It would seem therefore, that for all the alkali metals and alkaline earths save sodium, beryllium and magnesium, the reaction:

$$M(s) \rightarrow M^{n+}(aq) + ne^-(aq) \tag{4.39}$$

is thermodynamically feasible at 25°. According to one report, hydrated electrons can be prepared in this way, but more recent work suggests that they are only barely observable, short-lived intermediates in the overall hydrogen evolution reaction. Indeed, whatever the method of preparation, the hydrogen overvoltage in water seems too small for hydrated electrons to have more than a transient existence.

4.12. The oxidation state diagram. A convenient way of presenting the electrode potentials involving a particular element is shown in fig. 4.5. Beginning with the element in its standard state at the origin, the standard electrode potentials of the couples involving it and another oxidation state are multiplied by the oxidation state of the element in the other component. These values are then plotted as the ordinate, against the oxidation states as abscissae. The properties of the resulting diagram are as follows:

(*a*) The gradient of the line joining the values for any two oxidation states is equal to the potential of the couple containing them.

(*b*) If the point corresponding to the intermediate of any three oxidation states lies above the line joining the values for the highest and lowest, then the intermediate oxidation state is unstable to disproportionation into the other two. On the other hand, if the point lies below the line, then the intermediate oxidation state is stable to this reaction. This follows from (*a*) above and the discussion in the first part of §4.8.

Figure 4.5 depicts the oxidation state diagram for sulphur in acid and alkaline solution. The line gradients corresponding to the potentials of the oxygen and hydrogen electrodes and the iodine–iodide couple under these conditions are also shown.

Whereas in acid solution, thiosulphate is unstable to disproportionation into sulphur and sulphite, in alkali the converse is true. Although the reaction:

$$S_2O_3^{2-} \rightarrow SO_3^{2-} + S \qquad\qquad [4.40]$$

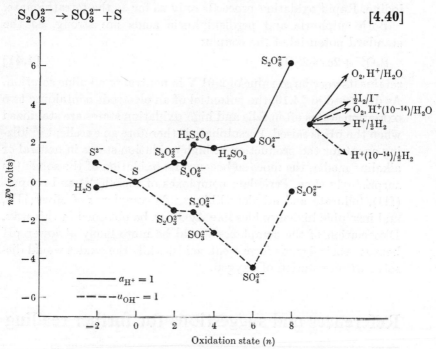

Fig. 4.5. The oxidation state diagram for sulphur in acid and alkaline solution.

does not involve hydrogen ions, sulphite is stabilized in acid solution because sulphurous acid is much weaker than thiosulphuric. Thus thiosulphate can be prepared by the dissolution of sulphur in an alkaline solution of a sulphite, but if a solution of sodium thiosulphate is acidified with dilute hydrochloric acid, sulphur and sulphurous acid are formed.

The powerfully reducing dithionites disproportionate to the same products in acid solution, but in alkali, thiosulphate and sulphite are formed. They are also oxidized to sulphite by dissolved oxygen at all acidities.

The disproportionation of sulphur is thermodynamically feasible in alkaline solution, but too slow to be observed. This may be due to strong S—S bonds in the elemental form, for the heat of atomization of sulphur is over 60 kcal/g-atom. Another example of kinetic inhibition is the conversion of thiosulphate to sulphate by iodine. Rapid oxidation proceeds only as far as the tetrathionate.

Both sulphuric and perdisulphuric acids are strong, so the standard potential of the couple:

$$S_2O_8^{2-} + 2e \rightarrow 2SO_4^{2-} \qquad\qquad [4.41]$$

retains its very large value of 2·01 V in neutral or alkaline solution. As discussed in §4.10, the potential of an electrode containing two oxidation states often falls and high oxidation states are stabilized when the pH is raised. Persulphate is therefore an excellent oxidizing agent for the production of high oxidation states in neutral or alkaline media, the more so because its oxidation of the solvent is surprisingly slow. Periodate complexes of nickel(IV) and copper-(III), tellurate and ethylenedibiguanide complexes of silver(III), and insoluble higher oxides like AgO can be obtained in this way. Dissociation of the complexes would be more likely at lower pH because their ligands form weak acids, while the oxides would dissolve with evolution of oxygen.

References and suggestions for further reading

Hunt, J. P. (1963). *Metal Ions in Aqueous Solution*, pp. 38–41, New York: W. A. Benjamin.

Jolly, W. L. (1956). *J. Chem. Educ.*, **33**, 512. (An account of the determination and use of electrode potentials in liquid ammonia.)

Kolthoff, I. M. and Belcher, R. (1957). *Volumetric Analysis*, vol. III. New York: Interscience.

Latimer, W. M. (1952). *The Oxidation States of the Elements and their Potentials in Aqueous Solution*, 2nd edition. N.Y.: Prentice-Hall. (A unique compilation of oxidation potentials with many valuable examples of their relation to inorganic chemistry.)

Phillips, C. S. G. and Williams, R. J. P. *Inorganic Chemistry*, vols. I and II, Oxford University Press, 1965 and 1966. (Numerous examples of the use of oxidation state diagrams.)

Powell, R. E. and Latimer, W. M. (1951). *J. Chem. Phys.*, **19**, 1139.

Sharpe, A. G. (1960). *Principles of Oxidation and Reduction*, R.I.C. Monographs, no. 2 (An excellent introductory account of the uses of electrode potentials.)

5. The Solubility of Ionic Salts

5.1. Introduction. The solubility product of a salt in terms of activities, K_s, is related to the standard free energy of solution, ΔG_s^0, by the equation

$$-\Delta G_s^0 = RT \ln K_s. \tag{5.1}$$

Converting to kcals/mole and logarithms to the base ten at 25°,

$$-\Delta G_s^0 = 1.36 \log K_s. \tag{5.2}$$

As a first approximation, the solubility product in terms of molalities may be equated to K_s, and if the products of the dissolution to which K_s refers are the only ones in a significant concentration, a very rough value for the solubility of a salt may be obtained from the solubility product which may in turn be calculated from ΔG_s^0 by using (5.1). A critical discussion of the relation between solubilities and solubility products has been given by Meites, Pode and Thomas (1966).

The logarithmic dependence of (5.2) implies that a change of a factor of ten in the solubility product corresponds to one of only 1·4 kcal/mole in the free energy of solution. One consequence of this sensitivity of K_s to energy changes is that the energetic difference between 'soluble' and 'insoluble' salts is a very small one. For example, the standard free energy of solution of potassium nitrate is less than 3 kcal/mole more negative than that of potassium perchlorate, yet the first is classified as soluble and the second as insoluble. Another is that any treatment of solubilities based on a theoretical interpretation of ΔG_s^0 is dangerously exposed to error when even quite small approximations are introduced.

The most revealing thermodynamic cycle yet proposed for the discussion of the solubilities of ionic salts is the one shown in fig. 5.1. Here ΔG_L^0 and ΔG_i^0 are the molal free energy changes associated with the sublimation of the lattice into the separate gaseous ions in the ideal gas state at one atmosphere pressure, and with the immersion of these ions in the solvent to form an infinitely dilute solution. Thus,

$$\Delta G_s^0 = \Delta G_L^0 + \Delta G_i^0. \tag{5.3}$$

In water, the two quantities on the right-hand side of this equation are of the order of some hundreds of kcal/mole but of opposite sign, while a typical free energy of solution lies between $+20$ and -20 kcal/mole. This discussion of ΔG_s^0 in terms of the small difference between two large quantities only increases the difficulties mentioned in the previous paragraph.

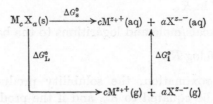

Fig. 5.1. Thermodynamic cycle for the dissolution of an ionic salt in water.

The major component of ΔG_L^0 is the lattice energy of the solid which is nearly always greater than 100 kcal/mole, so if the standard free energy of solution is to be negative, powerful forces must bind the ions and solvent molecules together in the final solution. Such forces readily arise in liquids which have high dielectric constants where strong interactions occur, to a first approximation, between the ions and the substantial dipole moments of the solvent molecules. Thus water, liquid sulphur dioxide and liquid ammonia dissolve many inorganic salts. When, as in solvents of low polarity, electrostatic interactions are very weak, the formation of solutions of inorganic salts is rare, and when it does occur, a different form of chemical binding in the solution must be invoked. Silver perchlorate dissolves in benzene but careful evaporation of the solution yields the compound $Ag(C_6H_6)ClO_4$ in which there are planar hexagons of carbon atoms. The C—C distances, although different from each other, are similar to those in benzene and the silver atoms lie above and below, but close to, the parallel benzene planes. Thus it seems likely that a strong interaction, similar to that found in ferrocene or $K[Pt(C_2H_4)Cl_3]$, operates in solution between the metal ion and the π-electrons of the solvent, and that this is what mainly offsets the high lattice energy of silver perchlorate when it dissolves in benzene.

5.2. The entropy of solution.

It is often true that, in a series of chemical reactions, the variations in the entropy contribution to the free energy changes are small compared with those in the enthalpy terms. Under these circumstances, the variations in ΔG within the series are determined by those in ΔH. The dissolution of ionic salts, where the enthalpy changes are usually comparatively small, is one type of process to which this assumption cannot be applied. This is particularly true of a series of salts which yield ions of differing charge in solution. According to empirical rules proposed by Latimer, the standard molal entropy of a solid compound is roughly equal to $\Sigma[\frac{3}{2}R \ln M - 0.94]$, where M is an atomic weight and the summation extends over all the atoms of the molecule (see page 9). This logarithmic dependence means that the entropies of solid salts vary comparatively little over the normally studied range of molecular weights. On the other band the entropies of ions fall off rapidly with decreasing size and, in particular, with increasing charge (see page 70). Consequently, entropies of solution become progressively more negative as the dissolved ions that are produced become smaller or more highly charged.

The entropies of solution of alkali metal salts with their large singly charged cations are rarely very negative and frequently positive. Many alkali metal salts dissolve with absorption of heat, but favourable entropy terms ensure that compounds of low solubility are rare. Thus the enthalpies of solution of KCl, KBr, KI, KNO_3 and K_2SO_4 are 4·2, 4·8, 4·9, 8·4 and 5·8 kcal/mole respectively, but the corresponding entropy charges are 18, 21, 26, 28 and 11 cal/deg.mole. Similarly, large singly-charged anions like perchlorate and nitrate have large positive entropies in water and this implies a favourable entropy of solution which is usually sufficient to ensure high solubility. All known simple nitrates are soluble and many perchlorates, like those of the dipositive transition metal ions, are highly deliquescent.

The carbonate ion is similar to the nitrate ion in size, but the double charge results in a much lower entropy. The entropies of solution of the compounds MCO_3, where a doubly charged cation is also produced, are about -40 cal/deg.mole and, although salts of this type have enthalpy terms which, by themselves, favour dissolution, they are invariably insoluble.

When the anion and cation that would be formed on dissociation are both triply charged, the entropy of solution is exceedingly negative. For lanthanum orthophosphate it is considerably less than -100 cal/deg.mole, and it is hardly surprising that the orthophosphates of scandium, yttrium and the lanthanides can be precipitated from solutions containing the tripositive ions of the metals. Indeed, apart from those which contain singly charged cations, phosphates, when they are prepared under conditions in which they are stable to hydrolysis, are invariably insoluble. The use of highly charged ions like Fe^{3+} and Sn^{4+} as phosphate separators in qualitative analysis is an important application of this property.

Even if an anion is only singly charged, provided it is small the compounds formed with triply charged cations will have very negative entropies of dissociation. The heats of solution of the hydroxides and fluorides of yttrium and the rare earths, by themselves, imply at least sparing solubility, but entropies of solution of about -80 cal/deg.mole result in solubility products of the order of 10^{-20}.

Despite the importance of the entropy of solution in determining solubility variations, there are many occasions when the enthalpy term has an equal or greater influence, and any theoretical treatment of solution energies must examine the free energy as a whole. It is possible to gain some insight into the variations in ΔG_s^0 for anhydrous salts by using semi-empirical expressions for ΔG_L^0 and ΔG_i^0 which are dependent on the crystal radii of the ions. The following sections describe the formulation of these expressions and their application to this problem.

5.3. The free energy of sublimation of the lattice.

For the compound M_cX_a, we may split up the free energy of the lattice into the corresponding enthalpy and entropy terms and use (2.14) for the lattice energy of the crystal:

$$\Delta G_L^0 = \Delta H_L^0 - T\Delta S_L^0, \tag{5.4}$$

$$= U[M_cX_a] + (a+c)RT - T\Delta S_L^0, \tag{5.5}$$

$$= \frac{256(a+c)z_+z_-}{r_+ + r_-} + (a+c)RT - T\Delta S_L^0. \tag{5.6}$$

The contribution of each atom to the molal entropy of a solid compound is given, to a good approximation by (1.20). Thus,

$$S^0_{298}[M_cX_a] = \tfrac{3}{2}R \ln M^c_M M^a_X - 0 \cdot 94(a+c),$$ (5.7)

where M_M and M_X are the atomic weights of M and X respectively.

The entropies of gaseous monatomic ions are given by the Sackur–Tetrode equation which, at 25° and one atmosphere pressure, reduces to,

$$S^0_{298} = \tfrac{3}{2}R \ln M + 25 \cdot 9,$$ (5.8)

where the small terms arising from the degeneracies of ground states have been neglected and the units are cal/deg.mole. Thus the molal entropy of the sublimed lattice is given by

$$S^0_{298} = \tfrac{3}{2}R \ln M^c_M M^a_X + 25 \cdot 9(a+c),$$ (5.9)

and

$$\Delta S^0_L = 26 \cdot 8(a+c).$$ (5.10)

Thus from (5.6) and (5.10), at 25°,

$$\Delta G^0_L = \frac{256(a+c)z_+ z_-}{r_+ + r_-} - 7 \cdot 4(a+c)$$ (5.11)

(5.11) has been derived by assuming that the gaseous ions formed by the sublimation of the lattice are monatomic.

5.4. The free energy of hydration of the ions. This property may be experimentally determined if experimental values of the standard enthalpies of formation of the gaseous ions are available. The molal entropies of M and X in their standard states are measured by the usual methods, and those of gaseous monatomic ions may be calculated from the Sackur–Tetrode equations. Using the alkali metal halides as an example, for the reaction,

$$M(s) + \tfrac{1}{2}X_2(\text{ref. state}) \rightarrow M^+(g) + X^-(g), \qquad [5.1]$$

$$\Delta G^0 = \Delta H^0_f[M^+(g)] + \Delta H^0_f[X^-(g)] - T(S^0[M^+(g)]$$
$$+ S^0[X^-(g)] - S^0[M] - \tfrac{1}{2}S^0[X_2]). \quad (5.12)$$

Now $\Delta G^0_f[M^+(aq)]$ and $\Delta G^0_f[X^-(aq)]$, the standard free energies of the reactions,

$$M(s) + H^+(aq) \rightarrow M^+(aq) + \tfrac{1}{2}H_2(g) \qquad [5.2]$$

and

$$\tfrac{1}{2}X_2(\text{ref. state}) + \tfrac{1}{2}H_2(g) \rightarrow X^-(aq) + H^+(aq) \qquad [5.3]$$

may be obtained from the appropriate electrode potentials. For the process,

$$M(s) + \tfrac{1}{2}X_2(\text{ref. state}) \rightarrow M^+(aq) + X^-(aq),\qquad\qquad [5.4]$$

$$\Delta G^0 = \Delta G_f^0[M^+(aq)] + \Delta G_f^0[X^-(aq)]\qquad\qquad\quad (5.13)$$

and ΔG_i^0 is the difference between the right-hand sides of (5.12) and (5.13). A set of specimen values for the alkali metal halides is shown in table 5.1.

TABLE 5.1   *Standard free energies of the reaction:*

$$M^+(g) + X^-(g) \rightarrow M^+(aq) + X^-(aq)$$

for the alkali metal halides at 25° (kcal/mole)

	Li	Na	K	Rb	Cs
F	−224·3	−200·5	−182·8	−177·7	−170·0
Cl	−195·6	−171·8	−154·1	−149·0	−141·3
Br	−188·7	−164·9	−147·2	−142·1	−134·4
I	−179·8	−156·0	−138·3	−133·2	−125·5

The precise constancy of the figures for, say, MF–MCl as M varies is contrived, because the values have effectively been obtained by using values of ΔH_f^0 or ΔG_f^0 for individual gaseous or aquated ions. Nevertheless, this constancy is still a reflection of a fact tested to the limits of experimental accuracy, namely that, in dilute solutions, the thermodynamic properties of one ion are independent of those of another. It therefore implies that the free energy of hydration can be split up into contributions from the individual ions. It is clear that if an absolute figure is assigned to the free energy of hydration of any one ion, all other values can, in principle, be derived from this, and that this set of values will still satisfy the additivity principle whatever the assigned figure may be. However, from the theoretical standpoint, it is important to make sure that these individual absolute values are as 'correct' as possible.

A number of attempts have been made to achieve this end, but all involve some assumption that cannot be thoroughly substantiated. Nevertheless most of them yield results which are in fairly close agreement, and the figures quoted in table 5.2 are those given by Noyes (1962).

TABLE 5.2 *The free energies of hydration of some ions (kcal/mole)*

Ion	$-\Delta G_h^0$	Ion	$-\Delta G_h^0$
Li^+	120·8	Be^{2+}	586·7[a]
Na^+	97·0	Mg^{2+}	452·9
K^+	79·3	Ca^{2+}	378·2
Rb^+	74·2	Sr^{2+}	343·5
Cs^+	66·5	Ba^{2+}	312·7
F^-	103·5[b]	Ra^{2+}	305
Cl^-	74·8	Sc^{3+}	937·2
Br^-	67·9	La^{3+}	777·5
I^-	59·0	Ga^{3+}	1102·1
S^{2-}	306·1	In^{3+}	978·6

[a] Corrected to a recent value, $\Delta G_f^0 [Be^{2+}] = -90\cdot0$.
[b] Corrected to the values given by the National Bureau of Standards Technical Note 270–1.

A possible theoretical interpretation of these values is based upon an approach first suggested by Born. If a sphere of radius r' has a charge of z units placed upon it, first in vacuo and then in a solvent of dielectric constant ϵ, the respective amounts of electrostatic work done are $z^2e^2/2r'$ and $z^2e^2/2r'\epsilon$. This suggests that the free energy of solvation, $\Delta G_{solv.}^0$ is given in kcal/mole by,

$$\Delta G_{solv.}^0 = -\frac{N_0 z^2 e^2}{2r'}\left(1 - \frac{1}{\epsilon}\right) + 1\cdot32, \qquad (5.14)$$

where the term $1\cdot32$ kcal/mole is the free energy change which occurs on the compression of one mole of gas, at 25° and one atmosphere pressure, to one litre.

One of several possible criticisms that can be made of this model is that the radius of an ion is unlikely to be the same in the gas and solvent phases. In the particular case of hydration, this is of no great significance because the dielectric constant is so large that the second term in the bracket is negligible compared with the first. It follows that the radius of the ion in the gas phase, whatever that means, is the value required for r' in (5.14). Since the crystal radius is a parameter designed to reproduce anion–cation distances in solid compounds where powerful interatomic forces operate, we might expect that it would differ substantially from the radius required to satisfy (5.14). This is indeed the case. Substitution of the known physical constants in (5.14) gives,

$$\Delta G_h^0 = -\frac{163 \cdot 9 z^2}{r'} + 1 \cdot 32 \qquad (5.15)$$

for the free energy of hydration of an individual ion. In fig. 5.2, the data in table 5.2 has been used to calculate $-163 \cdot 9 z^2 / (\Delta G_h^0 - 1 \cdot 32)$

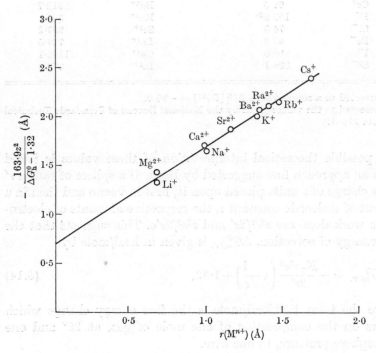

Fig. 5.2. Empirical variation of a function of the standard free energy of hydration of some ions with crystal radius.

for the alkali metal and alkaline earth cations. This has then been plotted against r, the crystal radius to be found in table 2.7. The resulting straight line has a gradient close to unity and indicates that, for cations, about $0 \cdot 7$ Å must be added to r to give r'. If the values of the constant are calculated for each separate cation and averaged, $0 \cdot 72$ Å is obtained with a standard deviation of $0 \cdot 01$ Å. This approach of an empirical fit to the Born Equation has been previously used by Latimer, Pitzer and Slansky (1939). For anions, the data is more meagre and less satisfactory, and for those included in table 5.2, r' exceeds r by between 0·3 and 0·6 Å.

There are other, more sophisticated ways of adjusting the Born equation to give an agreement with the experimental results. One particularly strong objection to (5.14) is that the macroscopic value of the dielectric constant is probably inappropriate in the high electric fields which operate close to the ions. A much smaller figure is more likely to be correct, and some treatments assume that r' is equal to the ionic crystal radius, and substitute for ϵ some suitable function of r and the macroscopic value of the dielectric constant. Nevertheless, all the methods are basically empirical, and we shall use the equation,

$$\Delta G_h^0 = -\frac{163 \cdot 9z_+^2}{r_+ + 0 \cdot 72} + 1 \cdot 3 \tag{5.16}$$

for the free energy of hydration of cations because it is simply related to the crystal radius. For practical purposes, (5.16) is not sufficiently accurate to justify the inclusion of the additional constant of 1·3 kcal/mole.

5.5. Variations in the free energy of solution with cation size and charge.
It is interesting to examine the alterations in the free energies of solution of a series of anhydrous salts of the same anion and formula type as the cation size is varied. Combining (5.3), (5.11) and (5.16) at 25° for a compound M_cX_a,

$$\Delta G_s^0 = \frac{256(a+c)z_+z_-}{r_+ + r_-} - \frac{164z_+^2 c}{r_+ + 0 \cdot 72} + a\Delta G_h^0 [X^{z-}] - 7 \cdot 4a - 6 \cdot 1c, \tag{5.17}$$

$$\frac{d(\Delta G_s^0)}{dr_+} = -\frac{256(a+c)z_+z_-}{[r_+ + r_-]^2} + \frac{164cz_+^2}{[r_+ + 0 \cdot 72]^2}. \tag{5.18}$$

When $d(\Delta G_s^0)/dr_+ = 0$,

$$\frac{[r_+ + r_-]^2}{[r_+ + 0 \cdot 72]^2} = \frac{256}{164} \cdot \frac{z_+z_-(a+c)}{cz_+^2}. \tag{5.19}$$

Using the relation $z_+c = z_-a$, taking the square root and neglecting the negative portion which has no relevance to this problem,

$$\frac{r_+ + r_-}{r_+ + 0 \cdot 72} = 1 \cdot 25 \sqrt{\left(1 + \frac{z_-}{z_+}\right)}. \tag{5.20}$$

It is fairly easy to show by further differentiation of (5.18) that (5.20) refers to a maximum. Thus a maximum in the free energy of solution occurs when,

$$r_- = r_+ \left[1{\cdot}25 \sqrt{\left(1 + \frac{z_-}{z_+}\right)} - 1 \right] + 0{\cdot}90 \sqrt{\left(1 + \frac{z_-}{z_+}\right)}. \qquad (5.21)$$

Although (5.17) has been derived only for salts of monatomic ions, (5.21) is as valid for those containing monatomic cations and a fixed polyatomic anion if the latter still makes an individual contribution to the entropy of the solid as the cation changes. Although the entropy of the gaseous polyatomic anion is larger than (5.8) suggests, the difference between it and the contribution to the entropy of the solid will still be a constant. On differentiation, this constant will naturally vanish.

Before discussing the implications of (5.21), it is worth making a few more comments on the relationship between the solubility of a compound and its free energy of solution. In comparing the solubilities of anhydrous salts of the same anion and formula type, it is not often that variations in activity coefficients with the molalities of saturated solutions will be enough to offset those in the standard free energies of solution. Consequently the order of the values of $-\Delta G_s^0$ will be the same as that of the solubilities of the anhydrous salts. Unfortunately the equilibrium between an anhydrous salt and its saturated solution is sometimes unattainable because full dissolution is blocked by the separation of a solid hydrate. Among the alkaline earth hydroxides for example, the octahydrates are the phases in equilibrium with saturated solutions of the strontium and barium compounds at 25°, while in the corresponding sulphate series, the tetrahydrate, heptahydrate and dihydrate are the compounds deposited from beryllium, magnesium and calcium sulphate solutions. In both these series, the measured molalities of the saturated solutions follow the order of the value of $-\Delta G_s^0$ [anhydrous salt] because the dissolution of the unhydrated compound proceeds far enough towards equilibrium before the precipitation of the hydrate begins. Sometimes however, the correlation between the two sequences may be destroyed by hydrate formation.

5.6. Some implications of §5.5. Because (2.14) and (5.16) are very approximate, we cannot expect (5.21) to predict accurately

TABLE 5.3   *The free energies of solution of some alkali metal and alkaline earth salts at 25° (kcal/mole)*

	Li⁺	Na⁺	K⁺	Rb⁺	Cs⁺	Be²⁺(a)	Mg²⁺	Ca²⁺	Sr²⁺	Ba²⁺
F⁻	+3·3	+0·6	−6·1	−9·2	−14·0	v. sol.	+9·6	+13·4	+12·4	+6·3
Cl⁻	−9·9	−2·2	−1·2	−2·0	−2·2	−41·8	−30·1	−15·6	−9·2	−2·9
Br⁻	−13·6	−4·1	−1·4	−1·3	−0·4	−55·4	−38·8	−24·5	−16·0	−8·7
I⁻	−18·6	−7·3(b, c)	−2·8	−2·0	−0·1	−64·6	−47·7	−30·3	−22·9	−15·6
OH⁻	−1·9	−10·1	−15·6	−18·0	−20·1	+28·9	+15·1	+6·9	−0·5	−4·5
NO₂⁻	+0·9	−3·0	−8·4	—	−8·8	—	—	−4·5	−4·6	−11·1
HCO₃⁻	—	+0·7	−2·1	−3·4	−0·1	—	—	—	—	—
NO₃⁻	−3·5	−1·6	+0·1	−0·6	−0·1	—	−21·2	−7·7	−0·6	3·1
ClO₄⁻	v. sol.	−3·8	+2·7	+3·2	+3·3	—	−34·5	—	—	−11·0
S²⁻	—	−16·5	−29·0	−32·2	−35·1	−12·9	−3·3	+4·0	−4·1(c)	−7·4
CO₃²⁻	4·0	−1·0	−8·7(c)	−11·2	−17·4	—	+10·8	+11·4	+12·5	+12·0
SO₄²⁻	−2·4(c)	+0·3	+2·4	+0·4	−1·5	−8·0	−5·8	+6·1	+8·3	+12·1

(a) Corrected to ΔG_f^0 [Be²⁺(aq)] = −90·0 kcal/mole.
(b) Corrected for an error in the Circular 500 value for ΔG_f^0 [NaI(aq)].
(c) Corrected for an error in Latimer's calculated value of ΔG_f^0 for the solid.

the value of the cation radius at which a maximum in the standard free energy of solution is achieved in a series of salts of the same anion and formula type. Nevertheless, its implications with regard

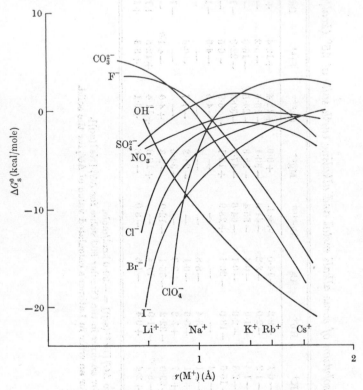

Fig. 5.3. Variations in the standard free energies of solution of some alkali metal salts with the crystal radius of the cation.

to qualitative variations in ΔG_s^0 are surprisingly vindicated. For an anion of minimum size 1·3 Å, these are as follows:

(a) For salts of the same anion and formula type, if r_+ is increased from $-0·72$ Å, the standard free energy of solution rises, reaches a maximum and then falls steadily towards a limiting value of $(a\Delta G_h^0[\mathrm{X}^{z-}] - 7·4a - 6·1c)$. Since ionic radii, in common experience, are restricted to values between about 0·5 and 3 Å, the series of experimentally observed free energies may only cover the increasing or decreasing portion of the curve, but in fortunate cases a maximum might be observed. If z_-/z_+ is the same for two series, each containing a fixed anion, ΔG_s^0 reaches a maximum at a larger cation radius in the series with the bigger anion. The maximum should occur when the sizes of the cation and anion are suitably matched.

(b) If the size of the cation in a salt of a particular anion and formula type is increased, a fall in the free energy of solution should never be succeeded by a rise.

(c) When two series of compounds have different formula types, but contain fixed anions of similar size, the free energy maximum should occur at a smaller cation radius in the series with the higher value of z_-/z_+.

These generalizations may be tested with the aid of the data in table 5.3 which contains the standard free energies of solution of over 100 inorganic salts. The values are mainly from Latimer (1952), and many depend on estimated entropies of solids. The figures for some alkali metal compounds are plotted against cation radius in fig. 5.3. The curves obtained are a convincing confirmation of generalizations (a) and (b). In the subsequent sections, extensive use is made of the ionic radii in tables 2.7 and 2.8, and the solubility of a compound is taken to be the molality of the solution in equilibrium with the most stable solid phase that can be obtained from a saturated solution.

5.7. Salts of the type MX. According to (5.21) the ions F^-, OH^-, NO_2^- and HCO_3^- are too small for the free energy maximum to be attained within the range of the crystal radii of the alkali metal ions (0.6–1.7 Å). Table 5.3 confirms this prediction, for the standard free energies of solution of the alkali metal salts of these anions become more negative as the series is descended. Hydrates of lithium, sodium and potassium hydroxides and of potassium, rubidium and caesium fluorides are formed, but the solubility sequence down these series departs from that in $-\Delta G_s^0$ only in moving from NaOH to KOH.

For larger anions like chloride and nitrate, ΔG_s^0 reaches a maximum at potassium within the range of cation radius studied. Hydrates of lithium chloride and nitrate are formed, but the solubilities correlate with the order in $-\Delta G_s^0$ except in moving from lithium to sodium nitrate. Other compounds of the type MX where

X^{z-} is moderately large are the carbonates and sulphides of the alkaline earths. Strontium is the least soluble carbonate, and in the sulphide series a maximum in the free energies of solution is observed at calcium.

When the anion in the compounds M^+X^- becomes larger, the maximum is attained at a cation radius greater than that of Cs^+. This appears to be the case for the alkali metal iodides and perchlorates. A further increase in $r(M^+)$ causes increasing solubility. Unlike the caesium and tetramethylammonium salts, tetraethylammonium perchlorate is not precipitated from solutions of the cation by perchloric acid. This agrees with the theory even though the cation is now polyatomic (but see page 112). In the alkaline earth series, the sulphates are an example where, even as far as radium, only the increasing portion of the free energy curve is observed.

There are two points where our predictions fail: in moving from potassium to caesium, the alkali metal bromides and nitrates show a decrease in ΔG_S^0 followed by an increase. In both series, the solubilities follow the sequence Na > K < Rb > Cs.

5.8. Salts of the type M_2X.

Equation (5.21) implies that the raising of the anion to cation charge ratio favours the attainment of the maximum free energy of solution at a smaller value of the cation radius. The sulphate and perchlorate ions are of comparable size but whereas the alkali metal perchlorates reach the solubility minimum at a cation radius greater than that of caesium, the sulphates attain theirs at potassium. A further satisfactory comparison can be made between the alkali metal carbonates and chlorides. Potassium has the least soluble chloride, but in spite of the fact that the carbonate anion has a slightly greater radius, the raising of the anion–cation charge ratio confines the alkali metal cation range to only the decreasing portion of the curve when the free energy of solution is plotted against the cation radius.

5.9. Salts of the type MX_2.

Relative to compounds MX, the low anion–cation charge ratio in these salts should cause larger values of the cation radius to be observed at the maxima in the standard free energies of solution. This prediction can be tested by comparing values for the sulphides and carbonates of the alkaline earth metals with those of the chlorides. All these anions have similar sizes, strontium forms the least soluble carbonate and the free-energy maximum is attained at calcium in the sulphide series, but for the chlorides, values of ΔG_s^0 increase steadily down the group. In saturated solutions of the chlorides, the equilibrium phase is a hydrate for all the metals including radium, but the solubilities follow the order in $-\Delta G_s^0$ (anhydrous salt), except in the step from magnesium to calcium.

The free energies of solution of the alkali metal fluorides diminish from lithium to caesium, but calcium forms the least soluble alkaline earth fluoride, so a free energy maximum must occur at about 1 Å. Again, the fact that potassium is the alkali metal chloride of lowest solubility contrasts with the steady increase in ΔG_s^0 down the corresponding alkaline earth series.

5.10. Variations in the free energy of solution with anion size and charge. On page 105 a rather large variation in the difference between the ionic radius and r' was noted, even for a very small number of monatomic anions. Furthermore, for poly-atomic anions, the molal entropy in the gas phase includes vibrational and rotational terms, so the difference between it and the contribution of the anion to the entropy of the solid is unlikely to be nearly constant from ion to ion. These facts reduce the chances of a successful treatment of the variations in the free energies of solution of salts of the same cation and formula type as the anion changes. If these sources of error are ignored and the problem is approached by differentiating the difference between lattice energy and ΔG_h^0 expressions similar to those used in §5.5, the conclusion obtained is that the general shape of the plot of ΔG_s^0 against anion radius is the same as those in fig. 5.3, that an increase in the cation to anion charge ratio favours, for cations of comparable size, an earlier occurrence of the free energy maximum, and that for cations of greater size in salts of the same formula type, this maximum is displaced to larger values of the anion radius.

Thus the free energies of solution of the lithium and sodium halides fall steadily from fluorine to iodine, KCl and RbBr are the least soluble of the potassium and rubidium halides, and the solubilities of the caesium salts diminish down the halide series. In the case of the alkaline earth compounds, a higher cation to anion charge ratio causes a fall in free energies of solution from the fluoride to the iodide for all five metals.

With polyatomic anions some success is still obtained. In the vertical series from OH^- to ClO_4^- in table 5.3, the anions increase in size. It appears that free energy maxima in the lithium and sodium columns occur at nitrite and bicarbonate respectively, while a steady increase is the rule for the three other alkali metals. However, series containing monatomic and polyatomic anions show no general agreement with the theory, probably for the reasons discussed at the beginning of the section.

5.11. Summary of sections 5.5–5.10. As far as one can tell from the data in table 5.3, the treatment of the variations in ΔG_s^0 with cation size is largely successful in describing overall tendencies.

It implies that a salt should be very soluble when there is a severe mismatch in cation and anion size. The presence of large ions means that lattice energies are low, especially if anion–anion or cation–cation contact occurs. The standard free energy of hydration of the small ion is then, by itself, nearly sufficient to ensure dissolution. 100 g of a saturated solution of lithium chlorate contains 82 g of the salt at 25°. Again, tetramethylammonium fluoride is extremely deliquescent. With salts of this type, the tendency to dissolve is so great that they are sometimes soluble in liquids of lower dielectric constant. Anhydrous sodium and magnesium perchlorates are both very soluble in acetone, while lithium iodide dissolves in ether, a fact sometimes erroneously cited as evidence of covalent character in the solid.

In view of these facts, the well-known precipitating ability of the very large tetraphenylarsonium ion is an anomaly that does not fit the theory introduced in §5.5. The relative solubilities of sodium, caesium and tetraethylammonium perchlorate for example, suggest that $As(C_6H_5)_4ClO_4$ should readily dissolve, but it may be precipitated from aqueous solution and used as a weighing form for the gravimetric analysis of perchlorate. Tetrabutyl-ammonium perchlorate is also insoluble. Such exceptions must be expected. As rotational and vibrational terms contribute to the entropy of a gaseous polyatomic cation (see page 111), and because (5.16) has been derived empirically for only a few monatomic ions, it would be surprising if a single solubility law should, in general, embrace the salts of both monatomic, and very large polyatomic cations. As yet the severity of the deviations cannot be assessed because there is insufficient data on the solubilities and thermo-dynamics of tetra-alkylammonium salts. However, there is grow-ing evidence that the larger organic cations like $N(But)_4^+$ exert a 'structure-forming' effect on water which tends to enhance its ice-like quality (Frank & Wen, 1957). As most inorganic ions are struc-ture breakers, it is possible that the entropies of these large cations and hence the entropies of solution of salts containing them, are much lower than expected. This might account for the low solu-bilities of many of their salts.

The recognition of a solubility minimum in a series of salts of a fixed anion or cation at a particular cation or anion size can be important in preparative chemistry. The ion $RhCl_6^{2-}$ is reduced very quickly in solution via the reaction,

$$RhCl_6^{2-} + H_2O \rightarrow RhCl_5 . H_2O^{2-} + \tfrac{1}{2}Cl_2, \qquad [5.5]$$

but it may be plucked out of the aqueous phase and stabilized if its alkali metal salt is sufficiently insoluble. Attempts to prepare the potassium, rubidium, tetraethylammonium and tetrabutyl-ammonium salts have failed, but oxidation of Cs_3RhCl_6 or $[N(CH_3)_4]_3RhCl_6$ with ceric ions yields the insoluble green hexa-chlororhodate(IV) salts.

The conclusions reached in §5.6 do not apply very well to the salts of metals like silver and mercury whose properties differ markedly from those expected of an ionic compound. In these cases, there are contributions to the binding over and above those predicted by the simple equations for the lattice and hydration energies that are applicable in the alkali metal and alkaline earth series. Under these circumstances, it appears that the additional contribution to the lattice energy is usually predominant, and solubilities are lower than expected. The internuclear distance in mercuric fluoride lies between those in the corresponding calcium and strontium salts and all three compounds have the fluorite structure. However, the free energies of solution of the mercuric halides increase sharply down the series, while those of the calcium and strontium halides diminish. This contrast may be attributed to departures from models represented by lattice energy equations which embody ionic radii. These departures increase particularly severely down the mercuric halide series (see page 27). Although it is true that mercuric chloride, bromide and iodide are slightly soluble in water, this is due to the formation of molecular species in solution [Latimer (1952), page 181]. Their solubility products, $a_{Hg^{2+}} \cdot a_{X^-}^2$, are 10^{-14}, 10^{-19} and 10^{-28} respectively, in contrast to the ready solubility of the fluoride. The low solubility products of the three heavier mercuric halides are due to the large additional contributions to their lattice energies.

5.12. Hydrate formation. Although the solubility of a salt is often determined by the equilibrium between the solution and the anhydrous compound, the phase deposited from the saturated solution is frequently a hydrate. Under these circumstances, the thermodynamics of the reaction,

$$M_c X_a(s) + n H_2O(l) \rightarrow M_c X_a \cdot n H_2O(s) \qquad [5.6]$$

are of central interest.

Latimer (1952) showed empirically that the individual additive contribution of each mole of lattice water to the molal entropy of a solid hydrate is about 9·4 cal/deg.mole. If the anions and cations in the hydrate and anhydrous salts make equal additive contributions to the molal entropies of both, then as $S^0[H_2O(l)] = 16\cdot7$ cal/deg.-mole, ΔS^0 for [5.6] will be about $-7n$ cal/deg.mole. This unfavourable entropy change correlates with the restriction imposed on the water molecules when they are locked in the hydrate lattice. As a stable hydrate will only be formed if ΔG^0 is negative, [5.6] must be exothermic to the tune of at least $2n$ kcal/mole at 25°. Thus less heat is evolved by the dissolution of a hydrate than by that of an anhydrous salt to whose formation it is stable. Because the change from a more stable higher hydrate to a lower is an endothermic process, increasing temperature favours the precipitation of successively lower hydrates from a saturated solution. Thus in the temperature ranges -25 to $-18°$, -18 to $34°$, 34 to $37°$, 37 to $52°$ and 52 to $100°$, the solid phases in equilibrium with a saturated solution of zinc nitrate are the nonahydrate, the hexahydrate, the tetrahydrate, the dihydrate and the monohydrate respectively.

References and Suggestions for Further Reading

Frank, H. S. & Wen-Yang Wen (1957). *Discuss. Faraday Soc.* **24**, 133.

Latimer, W. M. (1952). *The Oxidation States of the Elements and their Potentials in Aqueous Solution*, 2nd edition, New York: Prentice Hall.

Latimer, W. M., Pitzer, K. S. & Slansky, C. M. (1939). *J. Chem. Phys.* **7**, 108.

Meites, L., Pode, J. S. F. & Thomas, H. C. (1966). *J. Chem. Educ.* **43**, 667.

Noyes, R. M. (1962). *J. Amer. Chem. Soc.* **84**, 513.

Phillips, C. S. G. & Williams, R. J. P. (1965). *Inorganic Chemistry*, vol. I, pp. 254–63. Oxford University Press. Includes a fairly detailed treatment of entropies of solution.

Roselaar, L. C. (1965). *Educ. Chem.* **2**, 135, 184 (errata). A useful discussion of entropies of solution.

Sharpe, A. G. (1964). *Educ. Chem.* **1**, 75. An account of the general phenomenon of solubility.

6. Transition Metal Chemistry and Related Topics

6.1. Introduction. The outstanding feature of transition metal chemistry is variation of oxidation state. Indeed, it is just this characteristic which gives each transition element its variety and interest. A thorough understanding of the property would require the solution of a number of specific problems. First there is the question of the relative stability of the different oxidation states of the same metal when it is complexed with the same ligand. That a number of such complexes exist is often a characteristic of transition elements that distinguishes them from other metals. In this connection, a few comments on the difficulties of the problem are made in §6.13. Other interesting questions are the variations in the relative stabilities of two complexes of the same metal in the same two oxidation states as the ligand changes, and of two compounds or complexes containing the metal in two different oxidation states down a vertical transition metal group. These two problems are not discussed in any detail for the reasons mentioned on page 147 and in §6.12. The main emphasis in the chapter is placed on the variations in the relative stabilities of two oxidation states of the same element, complexed with the same ligand, along a horizontal transition metal period. This problem has been selected because it is less intractable than the others that have been mentioned, and because it has important associations with the chemistry of the lanthanides and actinides. Before approaching it, the question of metal–ligand interaction is briefly discussed.

6.2. Ligand field theory. The intensive study of transition metal chemistry made during the last twenty years has been mainly due to the application of ligand field theories of bonding in complexes. These theories have shown that it is very instructive to separate from the total metal–ligand interaction, that of the ligands with the five d orbitals of the metal which, in a free atom or ion, are degenerate.

As an example, the results of the approach of six negatively charged ligands from equivalent directions are shown in fig. 6.1 (b). In this octahedral field, the d orbitals are raised in energy, but the d_{xy}, d_{yz} and d_{xz} or t_{2g} trio bear a symmetry relationship to the ligands that is different from that of the $d_{x^2-y^2}$ and d_{z^2} or e_g pair.

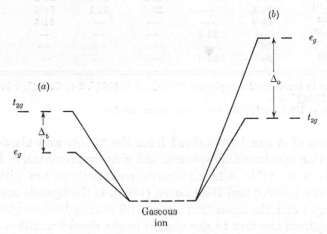

Fig. 6.1. Influence of (a) a tetrahedral, and (b) an octahedral ligand field on the d orbitals of a gaseous ion.

Thus, in the presence of the ligand field, the energies of the two sets are different, and it turns out that the t_{2g} orbitals are destabilized by less than the e_g.

Figure 6.1 (a) shows the effect of a tetrahedral field on the d orbitals of a free atom or ion. Δ, the splitting of the d orbitals, is less than in the octahedral case, partly because fewer ligands produce a weaker field. Theoretical calculations based on point charge models for metal and ligands suggest that, for the same ligand at the same internuclear distance,

$$\Delta_t = \tfrac{4}{9}\Delta_o \tag{6.1}$$

and this is roughly confirmed by experiment. Details of the various theories may be found in a valuable article by Cotton (1964). In this book, the crystal field theory which is the simplest, has been used.

T A B L E 6.1.   *Values of Δ in some transition metal complexes*

Oxidation state	6Cl	6F	6H$_2$O	6NH$_3$	3en	6CN
Mn^{II}	7·5	8·4	8·5	—	10·1	—
Ni^{II}	7·2	7·3	8·5	10·8	11·5	—
Cr^{III}	13·8	15·2	17·4	21·6	21·9	26·7
Co^{III}	—	—	18·2	22·9	23·2	33·5
Rh^{III}	20·3	—	27·0	34·1	34·6	—
Ir^{III}	25·0	—	—	—	41·4	—
Mn^{IV}	—	21·8	—	—	—	—
Pd^{IV}	22	—	—	—	—	—
Pt^{IV}	29	33·0	—	—	—	—

Values in tetrahedral complexes: $MnBr_4^{2-}$, 2·0; CoI_4^{2-}, 2·8; $CoBr_4^{2-}$, 2·9; $CoCl_4^{2-}$, 3·7; $Co(NCS)_4^{2-}$, 4·9.
1 unit = 1000 cm^{-1}; data from Jorgensen (1962a).

Values of Δ can be obtained from the visible and ultra-violet absorption spectra of complexes, and a few representative figures are given in table 6.1. Extensive compilations are given by Jorgensen (1962a) and Ballhausen (1962). If the ligands are fixed, Δ changes with the transition metal, and usually becomes larger in moving from the first to the second to the third transition series, or when the oxidation state of the metal increases. With a fixed metal and oxidation state, the changes in Δ are quite reproducible:

$$CN^- > NO_2^- > phen > en > NH_3 > EDTA$$
$$> H_2O > F^- > Cl^- > Br^- > I^-.$$

This sequence of ligands is known as the spectrochemical series and is obviously not dependent on any electrostatic feature such as charge or size. Attempts have been made to explain certain qualitative variations in it by means of the molecular orbital theory of transition metal bonding, but until the relative importance of the electronic effects invoked can be individually expressed in energy units, they must remain rather speculative.

6.3. Ligand field stabilization energies.

In a series of octahedral metal complexes of the dipositive oxidation state, the electron configuration of the free ion changes from d^0 to d^{10} as one moves from Ca^{2+} to Zn^{2+} across the first transition series. In the free ions, the degeneracy of the d orbitals implies that the distribution of the d electrons is spherically symmetrical, but this is not the case in the complexes. From d^1 to d^3, the electrons will lie with parallel spins in the t_{2g} orbitals which point between the ligands. From d^4 to d^7 inclusive, there are two possible configurations in each case: d^4, $t_{2g}^3 e_g^1$ or t_{2g}^4; d^5, $t_{2g}^3 e_g^2$ or t_{2g}^5; d^6, $t_{2g}^4 e_g^2$ or t_{2g}^6 and d^7, $t_{2g}^5 e_g^2$ or $t_{2g}^6 e_g^1$. These are described as high-spin and low-spin respectively.

Which of these two configurations is adopted depends upon the balance of two energy terms which are discussed more fully in §6.11, but for the moment, high spin behaviour will be assumed.

Under these circumstances, it is a profitable exercise to consider the energy stabilization conferred on each configuration relative to a hypothetical complex containing a spherical ion in which all the d orbitals are equally occupied. Relating the discussion to fig. 6.1 (b), for d^n, the spherical ion contains $n/5$ electrons in each of the d orbitals. Passing to a real high-spin complex where there is a total of m electrons in the e_g orbitals, then $\left(\dfrac{2n}{5} - m\right)$ electrons have dropped from the e_g to the t_{2g} set. The resultant stabilization is equal to $\left(\dfrac{2n}{5} - m\right)\Delta$. In table 6.2, these stabilizations are presented for both octahedral and tetrahedral environments. They are zero for the d^0, d^5 and d^{10} cases because, even in a ligand field, these ions are spherical.

TABLE 6.2 *Ligand field stabilization energies in an octahedral and a tetrahedral environment*

Configuration	Stabilization in octahedral field	Stabilization in tetrahedral field
d^0	0	0
d^1	$\frac{2}{5}\Delta$	$\frac{3}{5}\Delta$
d^2	$\frac{4}{5}\Delta$	$\frac{6}{5}\Delta$
d^3	$\frac{6}{5}\Delta$	$\frac{4}{5}\Delta$
d^4	$\frac{3}{5}\Delta$	$\frac{2}{5}\Delta$
d^5	0	0
d^6	$\frac{2}{5}\Delta$	$\frac{3}{5}\Delta$
d^7	$\frac{4}{5}\Delta$	$\frac{6}{5}\Delta$
d^8	$\frac{6}{5}\Delta$	$\frac{4}{5}\Delta$
d^9	$\frac{3}{5}\Delta$	$\frac{2}{5}\Delta$
d^{10}	0	0

An experimental test of these arguments is obviously desirable. Figure 6.2 shows a thermodynamic cycle from which the standard enthalpy of the process,

$$M^{n+}(g) + nH^+(aq) + ne^-(g) \rightarrow M^{n+}(aq) + \frac{n}{2}H_2(g) \qquad [6.1]$$

at 25° can be derived.

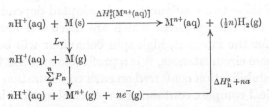

Fig. 6.2. Thermodynamic cycle for deriving variations in the standard enthalpies of hydration of similarly charged cations.

Here L_v is the latent heat of sublimation of the metal, $\sum_0^n P_n$ is the sum of the first n ionization enthalpies corrected from the ionization potentials to 25°, and ΔH_h^0 is the standard enthalpy of the process

$$M^{n+}(g) \rightarrow M^{n+}(aq). \qquad [6.2]$$

This is less than the enthalpy change associated with [6.1] by na, where a is the standard enthalpy of the reaction,

$$H^+(aq) + e^-(g) \rightarrow \tfrac{1}{2}H_2(g) \qquad [6.3]$$

and na is obviously a constant for ions of the same charge.

The experimental detection of ligand field stabilization energies is closely associated with the assumption that, because the reference state in a particular complex of particular charge contains a spherical ion, its total energy varies smoothly from metal to metal to across the transition series. This is not unreasonable, for every step is very similar to the one before, one proton being added to the nucleus and one fifth of an electron to each of the five d orbitals.

Now crystallographic studies of solid transition metal hydrates, followed by comparisons of solid and solution spectra, suggest that a number of the dipositive and tripositive aquated ions of the first transition series form octahedral hexa-aquo complexes in solution. In fig. 6.3, the octahedral ligand field stabilization energies in terms of spectroscopic Δ values have been subtracted from a plot of $-(\Delta H_h^0 + na)$ for the dipositive ions of the first transition series against atomic number. The resultant values do lie very close to a smooth curve through the d^0, d^5 and d^{10} configurations. Thus, provided the original assumption of the smooth variation of the hydration enthalpies of the spherical ions is accepted, fig. 6.3 pro-

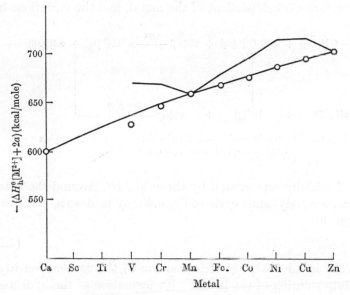

Fig. 6.3. Variations in the standard enthalpies of hydration of the dipositive ions of the first transition series.

vides a most convincing indication of the influence of ligand field stabilization energies on thermodynamic properties. The characteristic double hump with a cusp at d^5 is also observed in the lattice energies of the dihalides of the first transition series. Except for CaF_2, $ZnCl_2$, $ZnBr_2$ and ZnI_2, these compounds again contain six coordinated metal ions, although the octahedra are distorted by Jahn–Teller effects [see Orgel (1966)] in the d^4 and d^9 cases.

It is very important to remember that the ligand field stabilization energy usually comprises less than 10 per cent of the total energy released when the gaseous metal ion and ligands coalesce. An equivalent understanding of the remaining 90 per cent has not yet been achieved.

6.4. The stability constants of complexes. The formation of an octahedral transition metal complex in solution involves the replacement of water molecules by ligands:

$$M(H_2O)_6^{2+} + 6L \rightarrow ML_6^{2+} + 6H_2O. \qquad [6.4]$$

To a good approximation, the entropy change along the first transition series is independent of the metal, and the variations in

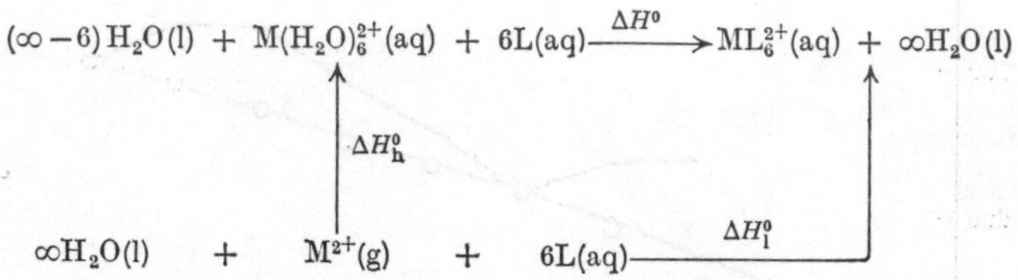

$$(\infty - 6)\,H_2O\,(l)\; +\; M(H_2O)_6^{2+}(aq)\; +\; 6L(aq) \xrightarrow{\Delta H^0} ML_6^{2+}(aq)\; +\; \infty H_2O\,(l)$$

$$\uparrow \Delta H_h^0 \qquad\qquad\qquad\qquad \uparrow \Delta H_l^0$$

$$\infty H_2O\,(l)\quad +\quad M^{2+}(g)\quad +\quad 6L(aq) \xrightarrow{\quad\quad} $$

Fig. 6.4. Thermodynamic cycle for the formation of a complex from an aquo-ion in aqueous solution.

ΔG^0 are faithfully represented by those in ΔH^0. Around the reaction, the thermodynamic cycle of fig. 6.4 may be drawn, and from this diagram:

$$\Delta H^0 = \Delta H_l^0 - \Delta H_h^0, \qquad (6.2)$$

where ΔH_l^0 is the enthalpy of immersion of the gaseous ion in a very dilute solution of the ligand with formation of the hydrated complex, ML_6^{2+}.

If both ΔH_l^0 and ΔH_h^0 are split up as ΔH_h^0 was in the previous section, then it is clear that, provided the differences in the spherical ion terms do not alter radically across the series, the variations in ΔH^0 will follow those in the difference between the ligand field stabilization energies. In fig. 6.5, log K for [6.4], less log K for the corresponding manganese system, is plotted for some complexes of oxidation state two from manganese to zinc. Assuming that Δ in each series does not vary greatly from metal to metal, the stabilities do in most cases follow the order of the difference between the ligand field stabilization energies for the complex and the aquo-ion: $Mn^{2+} < Fe^{2+} < Co^{2+} < Ni^{2+} > Cu^{2+} > Zn^{2+}$. The exceptions are found at Cu^{2+} which is presumably distorted from octahedral symmetry and stabilized by strong Jahn–Teller effects, and at $Fe(phen)_3^{2+}$, which is a low-spin complex (see §6.11). The usual sequence of stability constants for high-spin complexes, known as the Irving–Williams order, is in fact,

$$Mn^{2+} < Fe^{2+} < Co^{2+} < Ni^{2+} < Cu^{2+} > Zn^{2+}.$$

When the metal is fixed and the ligands altered, the ligand field contribution to ΔH^0 is still dominant and the order of stability follows the spectrochemical series; phen > en > EDTA > H_2O.

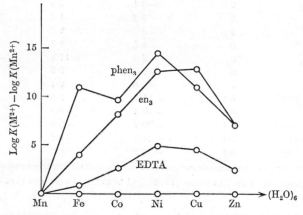

Fig. 6.5. Variations in the stability constants of some first row transition metal complexes, relative to manganese as zero.

The spread of values for Zn^{2+} is a measure of the variations in the differences between the spherical ion terms for ΔH_h^0 and ΔH_l^0 across the series. Relative to those in the differences between the ligand field stabilization energies these are fairly small, and this assumption is always necessary if the order of the stability constants is to be related to ligand field effects.

The enthalpies of the complexing reactions of transition metal ions are often numerically small, particularly with weak field ligands, and the absolute value of the stability constant may then be determined largely by the entropy term. Many complexes of the alkaline earth metals and the lanthanides also show this behaviour. According to (4.7), the entropies of ions are low when they are small and, in particular, when they are highly charged. Thus the combination of ions with high but opposite charges in a complexing reaction has a large positive entropy change. As the data in table 6.3 shows, in spite of generally unfavourable enthalpy terms,

TABLE 6.3 *Thermodynamic properties of some complexing reactions at 25°*

Reaction	Ionic strength	ΔG (kcal/mole)	ΔH (kcal/mole)	ΔS (cal/deg. mole)
$Ca^{2+} + SO_4^{2-}$	0	$-3\cdot2$	$1\cdot7$	16
$Mn^{2+} + SO_4^{2-}$	0	$-3\cdot1$	$3\cdot4$	22
$Co^{2+} + SO_4^{2-}$	0	$-3\cdot2$	$1\cdot7$	16
$Fe^{3+} + SO_4^{2-}$	0	$-5\cdot7$	$6\cdot2$	39
$Ce^{3+} + SO_4^{2-}$	0	$-4\cdot6$	$4\cdot7$	31
$Ce^{3+} + ClO_4^-$	0	$-2\cdot6$	$-11\cdot8$	-31
$Fe^{3+} + Cl^-$	0	$-2\cdot0$	$8\cdot5$	35
$Fe^{3+} + Br^-$	0	$-0\cdot8$	$6\cdot1$	23
$Mg^{2+} + EDTA^{4-}$	$0\cdot1$	$-11\cdot7$	$3\cdot5$	51
$Ba^{2+} + EDTA^{4-}$	$0\cdot1$	$-10\cdot4$	$-4\cdot9$	19
$La^{3+} + EDTA^{4-}$	$0\cdot1$	$-20\cdot7$	$-2\cdot9$	60
$Dy^{3+} + EDTA^{4-}$	$0\cdot1$	$-24\cdot3$	$-1\cdot2$	77
$Yb^{3+} + EDTA^{4-}$	$0\cdot1$	$-25\cdot9$	$-2\cdot3$	79

Figures are from Nancollas (1966) and Stability Constants (1964).

sulphate is a reasonably good complexing agent, particularly with the higher charged cations.

Although a thorough test is impossible because the appropriate thermodynamic data are not available, it seems that unfavourable entropy terms are mainly responsible for the usually feeble complexing powers of the large, singly charged nitrate and perchlorate anions (compare page 99). From the figures in table 6.3, $\Delta H^0 = 16\cdot5$ kcal/mole, $\Delta S^0 = +62$ cal/deg.mole and $\Delta G^0 = -2\cdot0$ kcal/mole for the reaction,

$$CeClO_4^{2+}(aq) + SO_4^{2-}(aq) \rightarrow CeSO_4^+(aq) + ClO_4^-(aq) \qquad [6.5]$$

so if it were not for the very positive entropy term, perchlorate would be a much stronger complexing agent than sulphate in this case.

Because of their low affinity for metal cations, perchlorates and nitrates are often used as inert supporting electrolytes for studying the reactions of the better complexing ligands. Some interesting chemical results have also been observed. When solutions of manganese (II) are electrolysed at a platinum anode in 6M perchloric or nitric acids, the precipitation of the dioxide is almost immediate. In 6M sulphuric acid however, substantial concentrations of manganese(III) may be obtained because the formation of a stable sulphate complex arrests the disproportionation into Mn^{II} and MnO_2. The relative complexing powers of the three anions were also mentioned on page 86.

Another example of the importance of the entropy of complexing is the 'chelate effect'. When the same number of coordination sites are attacked and occupied by a chelating agent and unidentate groups, the translational entropy of *more than one* ligand is lost is the latter case. Although rotational and vibrational entropy changes are also involved, the translational term appears to be the more important, and the additional stability of chelate complexes is attributed mainly to this entropy effect. The values, $\Delta S^0 = 20$ and 31 cal/deg.mole for the reactions,

$$Ni(NH_3)_6^{2+}(aq) + 3en(aq) \rightarrow Ni(en)_3^{2+}(aq) + 6NH_3(aq) \qquad [6.6]$$

and

$$Ni(NH_3)_6^{2+}(aq) + 2dien(aq) \rightarrow Ni(dien)_2^{2+}(aq) + 6NH_3(aq), \qquad [6.7]$$

where unidentate ligands are replaced by bidentate $(CH_2NH_2)_2$ and tridentate $NH(CH_2CH_2NH_2)_2$ are examples of the effect.

The favourable entropies of complexing that arise from the combined influence of chelation and partial neutralization of charge are particularly apparent in reactions of the ethylenediaminetetraacetate ligand, $(CH_2COO^-)_2NCH_2CH_2N(CH_2COO^-)_2$, usually abbreviated to EDTA. This has six possible coordination sites at the two nitrogens and four acetates, although only five are usually linked to the metal. The data in table 6.3 shows that the stabilities of the complexes formed by this ligand are in a large part due to very positive values of ΔS^0. Advantage is taken of this stability in the EDTA titrations used for the analysis of the alkaline earths, manganese, cobalt, lead, bismuth and a large number of other metals.

6.5. Octahedral and tetrahedral coordination.

If (6.1) is used to express Δ_o in terms of Δ_t in table 6.2, and the tetrahedral stabilization energies are subtracted from the octahedral ones, then ligand field effects alone suggest the following order of preference for tetrahedral coordination:

$$Mn^{2+} = Zn^{2+} > Fe^{2+} > Co^{2+} > Cu^{2+} \gg Ni^{2+}.$$

Generally speaking, the spherical ion terms appear to favour tetrahedral coordination in moving from Ca^{2+} to Zn^{2+}. Compounds or complexes containing tetrahedrally coordinated Ca^{2+} are hard to find, but in the chloride, bromide, iodide and chalcogenides of zinc, the cation is found in this type of environment. Many tetrahedrally coordinated zinc complexes such as $Zn(CN)_4^{2-}$ are also known. This tendency is imposed on the ligand field effect and perturbs the order, but the proximity of cobalt and nickel in the transition series, allied with their implied separation in the above sequence, suggests that the ligand field effect should be strongly reflected in their stereochemistry. This is indeed the case. If concentrated hydrochloric acid is added to solutions of $Co(H_2O)_6^{2+}$, the blue tetrahedral complex $CoCl_4^{2-}$ is readily formed. Similar complexes are formed with bromide, iodide and azide, while the compound $HgCo(NCS)_4$ is used both as a weighing form for cobalt and a calibrant for magnetic balances. In contrast, nickel forms relatively few tetrahedral complexes, and they tend to be unstable in aqueous solution to $Ni(H_2O)_6^{2+}$ or some other octahedral species. If a mixture of two moles of caesium chloride and one of cobalt chloride is melted, it turns dark blue and on cooling the compound Cs_2CoCl_4, which contains the tetrahedral ion $CoCl_4^{2-}$, remains. The corresponding nickel melt is purple due to tetrahedral $NiCl_4^{2-}$, but on freezing it reverts slowly to an orange solid containing one mole of caesium chloride and one of $CsNiCl_3$. In the latter compound the nickel is surrounded by an octahedron of chloride ions.

6.6. The variations in the electrode potential of the couple M^{3+}/M^{2+} across the first transition series.

Except for scandium and titanium, every metal of the first transition series forms a dipositive aquo-ion in water, while the corresponding tripositive complexes are known from scandium to cobalt. Reduction of the latter usually yields the dipositive ion because liberation of hydrogen is preferred to precipitation of the metal. This, allied with the frequent occurrence of oxidation states two and three in the first transition series, makes the standard potential $E^0[M^{3+}/M^{2+}]$ a very important quantity. The potential is related to the standard free energy change of the reaction

$$M^{3+}(aq) + \tfrac{1}{2}H_2(g) \rightarrow M^{2+}(aq) + H^+(aq) \qquad [6.8]$$

by the equation,

$$\Delta G_e^0 = -FE^0 \tag{6.3}$$

and the problem of the variations in ΔG_e^0 from metal to metal may be restated by means of the cycle in fig. 6.6. In this cycle, ΔH_e^0 is the standard enthalpy of the process [**6.8**] and is related to ΔG_e^0 by the equation,

$$\Delta G_e^0 = \Delta H_e^0 - T\Delta S_e^0. \tag{6.4}$$

$$M^{3+}(aq) + \tfrac{1}{2}H_2(g) \xrightarrow{\Delta H_e^0} M^{2+}(aq) + H^+(aq)$$

$$\Delta H_h^0[M^{3+}] \uparrow \qquad\qquad \Delta H_h^0[M^{2+}] \uparrow$$

$$M^{3+}(g) + \tfrac{1}{2}H_2(g) \xrightarrow{-P_3 - a} M^{2+}(g) + H^+(aq)$$

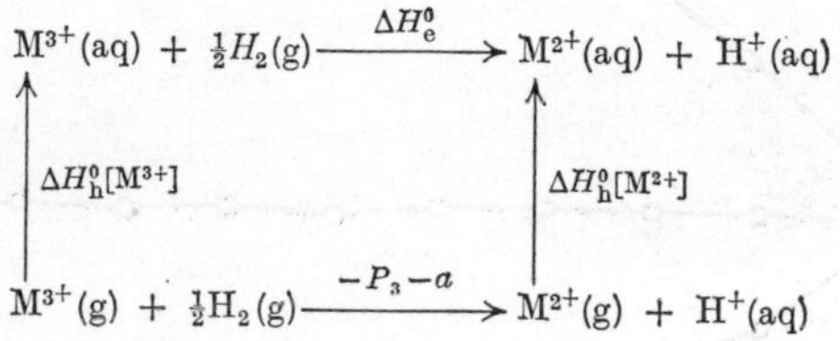

Fig. 6.6. Thermodynamic cycle for the process,
$$M^{3+}(aq) + \tfrac{1}{2}H_2(g) \to M^{2+}(aq) + H^+(aq).$$

From (6.4) and fig. 6.6

$$\Delta G_e^0 = \Delta H_h^0[M^{2+}] - \Delta H_h^0[M^{3+}] - P_3 - T\Delta S_e^0 - a, \tag{6.5}$$

where $\Delta H_h^0[M^{2+}]$, $\Delta H_h^0[M^{3+}]$, P_3 and a have the meanings defined on page 120.

Now $\Delta H_h^0[M^{2+}] + 2a$ and $\Delta H_h^0[M^{3+}] + 3a$ may be derived from fig. 6.2, and the additional constants, like that in (6.5), can be ignored, for we are interested only in the variations of these quantities across the transition series.

P_3 may be derived from the third ionization potentials of the metals and ΔS_e^0 can be found from the S^0 values for the individual species on each side of [**6.8**].

The relative contribution of each term on the right-hand side of (6.5) to the variations in ΔG_e^0 may now be assessed by reference to fig. 6.7. Here all the five varying quantities are plotted on the same scale against atomic number and, by a suitable adjustment of the energy zero which is different for each term, the graphs have been brought closer together for a convenient comparison. In table 6.4, the values of $E^0[M^{3+}/M^{2+}]$ are presented. The figures enclosed in brackets are only estimates, but their accuracy is sufficient to throw light on the reasons for the instability of one of the ions in their couples.

It is clear from fig. 6.7 that, within the framework of the restatement implied by (6.5), the following conclusions can be drawn:

(*a*) The entropy term $T\Delta S_e^0$ alters by a relatively small amount as one moves across the transition series, and the variations in ΔG_e^0

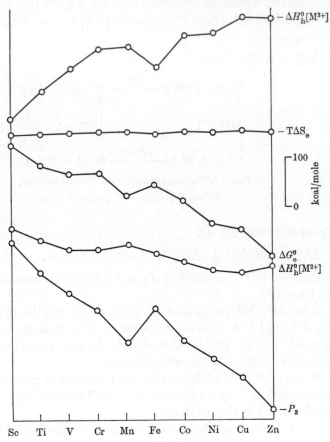

Fig. 6.7. Variations in the quantities in (6.5) across the
first transition series.

are therefore represented, to a good approximation, by those in ΔH_e^0.

(*b*) The changes in $-\Delta H_h^0[M^{3+}]$ are such that, taken alone, they would produce almost the exact opposite of the observed variation in ΔG_e^0, and they are only partially cancelled by the smaller fluctuations in $\Delta H_h^0[M^{2+}]$.

TABLE 6.4 $E^0[M^{3+}/M^{2+}]$ *for the metals of the first transition series at 25°*

Figures in brackets are estimates only

Metal	$E^0[M^{3+}/M^{2+}]$ (volts)	Metal	$E^0[M^{3+}/M]^{2+}$ (volts)
Sc	$(-2 \cdot 6)$	Fe	$0 \cdot 771$
Ti	$(-1 \cdot 1)$	Co	$1 \cdot 97$
V	$-0 \cdot 26$	Ni	$(4 \cdot 2)$
Cr	$-0 \cdot 41$	Cu	$(4 \cdot 6)$
Mn	$1 \cdot 60$	Zn	$(7 \cdot 0)$

(c) The adverse contribution of the term

$$(\Delta H_{\mathrm{h}}^0[M^{2+}] - \Delta H_{\mathrm{h}}^0[M^{3+}])$$

is overwhelmed by the comparatively large overall decrease in $-P_3$ which is only interrupted once by a sharp increase between manganese and iron.

(d) The importance of the term $-P_3$ can be seen from the fact that, as one proceeds along the transition series from metal to metal, the relative stabilities of the aquo-complexes parallel those of the gaseous ions except at chromium. Here, from fig. 6.7, chromium(III) in aqueous solution is more stable with respect to reduction to the dipositive state than either vanadium(III) or iron-(III) but, for the free ions in the gas phase, the opposite conclusion must be drawn. This solitary exception may be attributed to the occurrence of a large increase in $-\Delta H_{\mathrm{h}}^0[M^{3+}]$ between vanadium and chromium while $\Delta H_{\mathrm{h}}^0[M^{2+}]$ remains virtually unchanged.

Paragraphs (a) to (d) above, contain a restatement of the initial problem. A partial theoretical interpretation of this restatement is closely tied to one particular concept.

In an atom or ion, to a first approximation, any one electron may be regarded as moving in the field of a nucleus whose charge is screened by the others. For successive ions in a transition series, one proton is added to the nucleus and one electron to the d orbitals, but the additional screening effect of the diffuse d shell is insufficient to compensate for the increased nuclear charge. Consequently size parameters like ionic radii diminish across the series and, in accordance with Born-type equations like (5.15), the enthalpies and free energies of hydration of the spherical ions display an overall decrease. The changes in $\Delta H_{\text{h}}^0 [\text{M}^{3+}]$ are more violent than those in $\Delta H_{\text{h}}^0 [\text{M}^{2+}]$ because the two terms vary as z^2.

The cuspidal variations which are superposed on these decreases have been discussed in §6.3. The increase in ΔG_{e}^0 between vanadium and chromium may be attributed to the ligand field stabilization energies, for the tripositive state is stabilized by the addition of a third electron to the t_{2g} orbitals, while the dipositive complex is destabilized by the accommodation of a fourth in the e_g set. However, this is the only point where ligand field effects dominate the variations in ΔG_{e}^0 at the expense of $-P_3$.

Although the overall decreases in the ionic radii of similarly charged ions are slightly perturbed by ligand field effects, the changes from metal to metal are relatively small. Thus, in accordance with the Powell–Latimer equation and other arguments in §4.2, the entropy change, ΔS_{e}^0, remains fairly constant from scandium to zinc.

The most crucial problem is the interpretation of the changes in the third ionization potential which, within the framework of the chosen restatement, determine those in $E^0[\text{M}^{3+}/\text{M}^{2+}]$.

In any transition metal ion with the configuration $[\text{Ar}]\,3d^n$, the energy of the $3d^n$ shell may be broken down into two major components. The first is due to the coulombic attraction of the positively charged argon core for the d electrons, and the second is the result of interactions between these electrons. The second term may be further divided into two parts, one of which has a classical analogue and the other which has none. The classical part is the coulombic repulsion energy which operates between the like-charged electrons and is approximately proportional to the number of pairs. The non-classical part is called the exchange energy and exerts a stabilizing effect upon the ion. It is the energetic origin of Hund's Rule and is roughly proportional to the number of pairs of parallel spins. If the positive proportionality constants for the coulombic repulsion and exchange energies are J and K, and the attraction energy of each d electron in the field of the core is $-U$, then the energy, $E(d^n)$, of the d shell is approximately given by

$$E(d^n) = -nU + {}^n\text{C}_2 . J - mK, \tag{6.6}[1]$$

[1] This treatment of I_3 is taken from unpublished work by P. G. Nelson, Ph.D. thesis, Cambridge University (1962).

$^nC_2$ being the total number of pairs of electrons in the d shell, and m the total number of pairs of parallel spins. The ionization potential is then

$$E(d^{n-1}) - E(d^n) = U + J(^{n-1}C_2 - {}^nC_2) + K\delta m, \qquad (6.7)$$

$$= U - (n-1)J + K\delta m, \qquad (6.8)$$

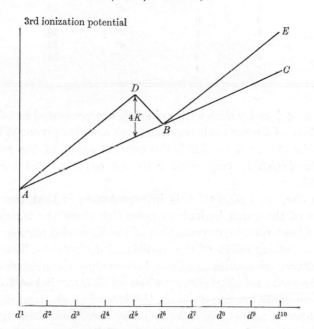

Fig. 6.8. Schematic variation of the third ionization potentials of the first row transition metals.

where δm is the decrease in the number of pairs of parallel spins. U and $(n-1)J$ should both increase smoothly with nuclear charge, so their difference may either increase or decrease. In fact, the experimental data show that the fall in the term $-(n-1)J$, which is a measure of the extent to which the ionizing electron is repelled by the remaining $(n-1)$, is outweighed by the rise in U, a reflection of the attraction of the increasing nuclear charge. Thus $U - (n-1)J$ increases smoothly with n and is represented by the curve ABC in fig. 6.8.

Upon this variation, the term $K\delta m$ is superimposed. Values of m, which are given by $(^xC_2 + {}^yC_2)$ where x is the number of electrons

TABLE 6.5   *The number of pairs of parallel spins for various* d^n *configurations and their decrease on ionization*

Con-figuration	m	δm	Con-figuration	m	δm
d^0	0	—	d^6	10	0
d^1	0	0	d^7	11	1
d^2	1	1	d^8	13	2
d^3	3	2	d^9	16	3
d^4	6	3	d^{10}	20	4
d^5	10	4	—	—	—

with spin $+\frac{1}{2}$ and y that with spin $-\frac{1}{2}$, are presented in table 6.5. The values of δm for each configuration are also given. When the term $K\delta m$ is added to ABC, the general shape of the resultant function $ADBE$ is very similar to the experimental ionization potential.

Thus the conclusion of this interpretation is that the overall increase of the third ionization potential along the series is the result of inadequate compensation of the increased nuclear charge by the screening effect of the additional d electrons. The crucial drop between manganese and iron is caused by the phenomenon of an unchanged exchange energy when an electron is lost from a d^6 ion. From configuration d^1 to d^5, the loss of an electron decreases the number of pairs of parallel spins by zero, one, two, three and four respectively, but at d^6 this decrease falls to zero again and the d^1 to d^5 pattern is repeated from d^6 to d^{10}.

The chemical consequences of the variation are apparent from table 6.4. At low values of n, the M^{2+} state is strongly reducing, and the estimated potentials for scandium and titanium imply that their unknown dipositive ions should decompose water. Thus the monoxide and dichloride of titanium instantly evolve hydrogen when added to dilute acid.

Aqueous solutions of vanadous and chromous ions are oxidized in air, but the increasing stability of the dipositive complex is very evident at manganese where Mn^{3+} is as powerful an oxidizing agent as permanganate. The fall in the third ionization potential between d^5 and d^6 may be neatly demonstrated by the oxidation of ferrous ions by manganese(III). Between iron and cobalt, the renewed increase in stability of the dipositive state is apparent, for the ion $Co(H_2O)_6^{3+}$ steadily oxidizes water, even though it is some-what stabilized by being a diamagnetic low-spin complex. The high potentials for nickel, copper and zinc suggest that the tri-positive ions are, in thermodynamic terms, highly unstable to reduction by water. When $Ba(CuO_2)_2 . 2H_2O$ and the hydrated oxides of nickel(III), prepared by hypochlorite oxidation and pre-cipitation in alkaline media, are dropped into acidified water, they are instantly reduced with evolution of oxygen.

The absence of quantitative comment leaves the theoretical interpretation in a rather unsatisfactory state. The discussions of the overall changes in ionization potential in terms of inadequate compensation of increased nuclear charge are no more than useful rationalizations of the observed results based on a simplified model of a many-electron atom. It is not certain that the qualitative dominance of nuclear charge could be proved from first principles, while the theoretical calculation of crystal field splittings, even with monatomic ligands, is notoriously difficult (Watson & Freeman, 1964).

This problem has been discussed at considerable length because, with understandable modifications, it can be extended to include several other important systems.

6.7. Dihalides and trihalides of the first transition series.

The instability of the trihalides of the first transition series is principally associated with the decomposition:

$$MX_3(s) \rightarrow MX_2(s) + \tfrac{1}{2}X_2. \tag{6.9}$$

The variations in the standard free energy change of this reaction may be treated by an extension of the problem of the previous section where, in (6.5), $\Delta H_h^0[M^{2+}]$ and $\Delta H_h^0[M^{3+}]$ are replaced by the standard enthalpies of the reactions:

$$M^{2+}(g) + 2X^-(g) \rightarrow MX_2(s) \tag{6.10}$$

and

$$M^{3+}(g) + 3X^-(g) \rightarrow MX_3(s). \tag{6.11}$$

The variations in these enthalpies are very similar to those in $\Delta H_h^0[M^{2+}]$ and $\Delta H_h^0[M^{3+}]$, and the qualitative conclusion of the treatment is that they serve only to mitigate the changes in the third ionization potential which is the term that mainly determines the variations in the standard free energy of [6.9] (Nelson & Sharpe, 1966).

For scandium, ΔG^0 is so positive that, at 25°, the unknown dichloride must disproportionate to the trichloride and the metal; $TiCl_2$, VCl_2 and $CrCl_2$ are stable to this reaction, while at manganese, the trichloride is improperly characterized and does not exist at room temperature due to ready loss of chlorine. The decrease is the ionization potential between manganese and iron is reflected in the marked stability of ferric chloride but, from cobalt to zinc, its renewed increase is apparent from the total instability of the unknown trichlorides.

Roughly speaking, substitution by fluorine adds a positive constant of about 60 kcal/mole to ΔG^0 for the chlorides. This is due to the easier formation of the third gaseous halide anion and to the increased difference in the higher lattice energies (see §2.8). The effect is to shift the range of metals with both halides characterized to higher atomic number. Thus the unknown ScF_2 and TiF_2 probably disproportionate like $ScCl_2$, while VF_2 is of borderline stability and the trifluorides of manganese and cobalt, but not those of nickel, copper or zinc are now stable at 25°.

The same stability variation is discernible in the chemistry of the tri-iodides, although they dissociate more readily than the three other types of halide. ScI_2 is unknown, but both iodides are well characterized for titanium, vanadium and chromium. The future preparation of MnI_3 is unlikely for it must be very unstable with respect to MnI_2, and, despite the fall in the third ionization potential between manganese and iron, appreciable quantities of ferric iodide may not be obtained at 25° due to a favourable dissociation into FeI_2 and iodine. The unknown triodides of cobalt, nickel, copper and zinc must also be highly unstable with respect to loss of iodine.

6.8. The oxidation states of the lanthanides.   Table 6.6 lists the properly characterized lanthanide halides, MX_2 and MX_4, together with the complex fluorides, A_2MF_6, the available electrode potentials of the couples M^{3+}/M^{2+} and M^{4+}/M^{3+}, and the configurations of the atoms. Figures enclosed in brackets are only estimates.

TABLE 6.6 *The dipositive and tetrapositive oxidation states of the lanthanides*

Element	Atomic configuration	Halides MX_2	$E^⦵[M^{3+}/M^{2+}]$ (volts)	+4 Halides and complex halides	$E^⦵[M^{4+}/M^{3+}]$ (volts)
La	$5d^16s^2$	—	—	—	—
Ce	$4f^26s^2$	—	—	$CeF_4, K_2CeF_6, Cs_2CeCl_6$	1·6
Pr	$4f^36s^2$	—	—	PrF_4 (a), K_2PrF_6 (b)	2·9(c)
Nd	$4f^46s^2$	$NdCl_2$, NdI_2	—	—	—
Pm	$4f^56s^2$	—	—	—	—
Sm	$4f^66s^2$	All halides	−1·55	—	—
Eu	$4f^76s^2$	All halides	−0·43	—	—
Gd	$4f^75d^16s^2$	—	—	—	—
Tb	$4f^96s^2$	—	—	TbF_4, K_2TbF_6 (b)	2·9(c)
Dy	$4f^{10}6s^2$	$DyCl_2$	—	(d)	—
Ho	$4f^{11}6s^2$	—	—	—	—
Er	$4f^{12}6s^2$	—	—	—	—
Tm	$4f^{13}6s^2$	TmI_2	−1·5(c)	—	—
Yb	$4f^{14}6s^2$	All halides	−1·15	—	—

(a) PrF_4 has not yet been prepared free of PrF_3.

(b) Fluorination of Cs_3NdF_6 and Cs_3DyF_6 yields products containing a considerable proportion of the metal in the +4 state.

(c) Estimates only.

(d) Attempts to fluorinate Cs_3HoF_6 have been unsuccessful.

The conducting diiodides like LaI_2 and CeI_2 are metal-like. They are usually written $M^{3+}(e^-)I_2$ and have been omitted.

Magnetic measurements made on the solid halides, together with the spectra of atoms and compounds, suggest that, whether combined or in the gas phase, the ions M^{3+} and M^{4+} have an electron configuration $4f^n$ with the $5d$ and $6s$ levels unoccupied. The electronic structures of the M^{2+} ions have not been fully elucidated, but it appears that in nearly all cases, the outer electrons are entirely in the $4f$ shell. This is true of those dipositive ions which occur in stoicheiometric salt-like compounds, and although the $5d$ and $4f$ orbitals are similar in energy at the beginning of the M^{2+} series where La^{2+} (g) has the structure $5d^1$, it is likely that, in any cases where ground states do not arise from the appropriate $4f^n$ configurations,[1] terms due to the latter are only just higher in energy. Consequently, for energetic purposes, it is probably a reasonable approximation to assume that the M^{2+} ions have $4f^n$ configurations.

A thermodynamic approach to the interesting problems offered by the oxidation states of the lanthanides and actinides is difficult because, at the present time, insufficient reliable data exists. In this section, and the one that follows, the available measurements are used to make cautious guesses at the trends in certain thermodynamic quantities across the two series. We emphasize that the discussion is speculative, and that future experiments may prove it partially incorrect.

Spectroscopic measurements on lanthanide compounds and complexes show that the crystal field splittings are only about two or three per cent of the values in the corresponding transition metal ions. Such observations suggest that the $4f$ orbitals are buried in the xenon core where they interact only weakly with ligand electrons. In addition, for ions of the same charge, the radii determined from oxides seem to decrease quite steadily across the series. This, combined with the presence of only small ligand field effects and equations like (2.14) and (5.16) suggests that thermodynamic properties such as lattice energies and hydration enthalpies should change fairly smoothly with atomic number.

Very few third or fourth ionization potentials of the lanthanides

---

[1] Apart from La^{2+} (g), $Gd^{2+}(4f^75d^1)$ seems to be the only other likely candidate in the gas phase. However, in chemical environments, ligand field effects would stabilize $4f^n5d^1$ configurations slightly.

TABLE 6.7 *The latent heats of sublimation of the lanthanides (kcal/mole) and the standard electrode potentials (volts) at 25°*

Metal	$L_v^{(a)}$	$E^0[M^{3+}/M]$	Metal	$L_v^{(a)}$	$E^0[M^{3+}/M]$
La	103	−2·52	Gd	96	−2·40
Ce	112	−2·48	Tb	94	−2·39
Pr	89	−2·47	Dy	71	−2·35
Nd	78	−2·44	Ho	71	−2·32
Pm	—	—	Er	82	−2·30
Sm	49	−2·41	Tm	59	−2·28
Eu	42	−2·41	Yb	36	−2·27

[a] Habermann and Daane (1964).

are known, but the steady fall in the radii of ions across the series suggests that they probably show an overall increase. This idea receives support from tentative I_3 values for lanthanum, cerium, praseodymium and ytterbium (see appendix 4). A further clue to the nature of the changes in I_3 is provided by the smooth change in the potential $E^0[M^{3+}/M]$ across the series,[1] (see table 6.7), an observation that is striking in view of the irregularities in the stabilities of the dipositive and tetrapositive oxidation states. This potential is given by $\Delta G^0/3F$, where ΔG^0 is the standard free energy of the process,

$$M(s) + 3H^+(aq) \rightarrow M^{3+}(aq) + \tfrac{3}{2}H_2(g). \qquad [6.12]$$

Ignoring small changes in ΔS^0 across the series, the variations in ΔG^0 from metal to metal are approximately given by those in the right-hand side of the equation,

$$\Delta G^0 = L_v + I_1 + I_2 + I_3 + \Delta H_h^0[M^{3+}] + \text{constant}. \qquad (6.9)$$

Tentative values of I_1 and I_2 are now available for all the lanthanide metals. Both quantities appear to change relatively little across the series, and the variations in $(L_v + I_1 + I_2)$ are determined almost entirely by those in the latent heats of sublimation given in table 6.7. L_v displays two rather perturbed, overall decreases from lanthanum to europium and from gadolinium to ytterbium,

[1] Values are based on rather unreliable heats of solution of the metals in acid. More recent, but less comprehensive data suggests that the changes, although less regular, are still *relatively* smooth.

separated by a sharp increase between europium and gadolinium. Now if the terms ΔG^0 and $\Delta H_h^0[M^{3+}]$ in (6.9) vary smoothly with atomic number while L_V falls, rises steeply and then falls again, the third ionization potential must show an equivalent drop between europium and gadolinium if it is to eliminate the irregularity in the

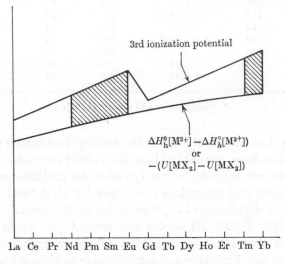

Regions with stable known dipositive states (except $DyCl_2$)

Fig. 6.9. Possible approximate variation in (a) the third ionization potential of the lanthandies and (b) the difference in the hydration enthalpies of the M^{2+} and M^{3+} ions or the lattice energies of halides MX_2 and MX_3.

heats and free energies of formation of the tripositive ions. Consequently there is strong evidence that, to a first approximation, I_3 displays the familiar saw-tooth variation[1] shown in fig. 6.9. The irregularity appears in the expected position for there is zero loss of exchange energy when an electron is lost from an f^8 ion.

The variations in the standard free energy of the reaction:

$$M^{3+}(aq) + \tfrac{1}{2}H_2(g) \rightarrow M^{2+}(aq) + H^+(aq) \qquad [6.13]$$

may now be treated as an extension of the problem in §6.6. In (6.5), $\Delta H_h^0[M^{2+}]$ and $\Delta H_h^0[M^{3+}]$ will no longer display severe cuspidal

---

[1] In view of the irregularities in L_V, considerable deviations from this idealized form of I_3 must be expected. This is taken up in more detail below.

variations but should vary fairly smoothly with atomic number. $\Delta H_{\mathrm{h}}^{0}[\mathrm{M}^{3+}]$ is more negative than $\Delta H_{\mathrm{h}}^{0}[\mathrm{M}^{2+}]$, and the changes in it, which we expect to be roughly proportional to z^2, will be more violent, so the term $(\Delta H_{\mathrm{h}}^{0}[\mathrm{M}^{2+}] - \Delta H_{\mathrm{h}}^{0}[\mathrm{M}^{3+}])$ will increase steadily across the series as the ionic radii decrease. If we assume that, as in §6.6, the overall increase in P_3 is more violent than that in $(\Delta H_{\mathrm{h}}^{0}[\mathrm{M}^{2+}] - \Delta H_{\mathrm{h}}^{0}[\mathrm{M}^{3+}])$, then the idealized conditions depicted in fig. 6.9 will prevail. From (6.5), the dipositive state is most stable to the tripositive when the ionization enthalpy exceeds the other term by the greatest amount, and the standard electrode potentials of the couples $\mathrm{M}^{3+}/\mathrm{M}^{2+}$ for samarium, europium, thulium and ytterbium are in accord with this prediction. The unknown dipositive ions should be even more unstable to oxidation by water than those that already exist. Since all of the latter save Eu^{2+} quickly reduce the solvent, the treatment suggests that it is very unlikely that any new dipositive ions will be prepared in aqueous solution.

The relative stabilities of the trihalides with respect to the process in [6.9] can also be discussed via fig. 6.9 if $(\Delta H_{\mathrm{h}}^{0}[\mathrm{M}^{2+}] - \Delta H_{\mathrm{h}}^{0}[\mathrm{M}^{3+}])$ is replaced by $-(U[\mathrm{MX}_2] - U[\mathrm{MX}_3])$. Again, the most stable dihalides are found in the shaded regions. This decomposition is relevant when one is considering the likelihood of the formation of dihalides by thermal decomposition of MX_3, a common method of preparing the di-iodides of samarium, europium and ytterbium, or by hydrogen reduction, the normal procedure used to obtain the remaining dihalides of these metals. As expected from fig. 6.9, formation of the dihalides of the other lanthanide metals is more difficult, and is achieved by the dissolution of the metals in the trihalide melts, followed by a rapid freezing of the equilibrium. Once made, the reverse reaction,

$$3\mathrm{MX}_2 \rightarrow \mathrm{M} + 2\mathrm{MX}_3 \qquad\qquad [6.14]$$

is the probable mode of decomposition of an unstable dihalide. The problem of the variations in the standard enthalpy of this decomposition in the solid state may be examined by means of the cycle in fig. 2.6 and equation (2.25):

$$\Delta H^0 = 3U[\mathrm{MX}_2] - 2U[\mathrm{MX}_3] + 2I_3 - (L_{\mathrm{v}} + I_1 + I_2). \qquad (2.25)$$

In this case, the relative stabilities of the dihalides are affected by changes in $(L_{\mathrm{v}} + I_1 + I_2)$ as well as by those in I_3.

It is interesting that the variations in the quantity $-(L_v + I_1 + I_2)$ in most cases serve only to supplement the postulated I_3 term, for, as we have seen, both display an overall increase across the series which is broken by a sharp fall between europium and gadolinium. Between dysprosium and erbium however, $-(L_v + I_1 + I_2)$ decreases considerably, and using the argument on pages 137–8 concerning the smooth variation in $E^0[M^{3+}/M]$, this suggests that the increase in I_3 is less severe in this region. In other words, in moving from gadolinium to ytterbium, small but significant downward and upward breaks occur at dysprosium and erbium respectively. This effect, which is not covered by the approximate treatment of the ionization potential given on page 130, seems to be connected with a stabilization conferred on an ion when the total angular momentum quantum number, L, of the ground state is large. According to Hund's rules, the ground term takes the maximum value of L consistent with the maximum value of S, and when $L = 6$ (I ground terms), this stabilization is particularly large. Moreover, it is more significant in the second half of the lanthanide series where the relevant parameter of interelectronic interaction is some 50 per cent greater than in the first. For Dy^{2+}, Ho^{2+} and Er^{2+}, the ground states are $^5I_8$, $^4I_{\frac{15}{2}}$ and $^3H_6$, while for Dy^{3+}, Ho^{3+} and Er^{3+}, they are $^6H_{\frac{15}{2}}$, $^5I_8$ and $^4I_{\frac{15}{2}}$ respectively. Thus Ho^{2+} is neutral to this effect, but Dy^{2+} is stabilized with respect to Dy^{3+} while Er^{2+} is destabilized with respect to Er^{3+}. This is believed to produce the suspected irregularity in I_3 and, together with the trend in L_v and the decrease in $(3U[MX_2] - 2U[MX_3])$, to be responsible for the unexpected stability of $DyCl_2$ and the absence of lower oxidation states of erbium from table 6.6 (careful attempts to prepare erbium dihalides by Er/ErX_3 melt reactions have proved unsuccessful). A treatment of the energies of $4f^n$ ground terms which includes these stabilizations has been given by Jorgensen (1962b and 1964) in connection with the charge transfer spectra of lanthanide complexes. The effect is significant in the lanthanide-(III)–lanthanide(II) problem because the changes in ΔH^0 for equation [6.14] across the series must be small, and so the order may be perturbed by relatively mild irregularities in I_3.

In contrast, at present it does not seem necessary to consider such second-order effects in the case of the tetrapositive oxidation states of the lanthanides. In a first approximation, the fourth ionization potential may be assumed to increase steadily with atomic number save for a sharp drop from gadolinium to terbium after the half-filled shell, and to determine the relative stabilities of the tripositive and tetrapositive oxidation states. The implied order of stability of the latter is then

$$\text{Ce} > \text{Pr} > \text{Nd} > \text{Pm} > \text{Sm} > \text{Eu} > \text{Gd} \ll \text{Tb} > \text{Dy} > \text{Ho}$$

etc. and, as table 6.6 shows, this sequence is consistent with the experimental facts.

In comparing corresponding $M^{II} \rightarrow M^{III}$ processes in the lanthanide and first transition series, we see that the general overall increases in ΔH^0 are more severe in the latter case and that, in the transition series, the ligand field stabilization energies in $-\Delta H_h^0[M^{3+}]$ or $U[MX_3]$ (see fig. 6.7) greatly reduce the influence of the drop in the third ionization potential on ΔH^0. These two effects combine to produce some interesting differences between the two types of problem. Thus $E^0[Mn^{3+}/Mn^{2+}] = 1.6$ V, but despite a fall of over 70 kcals/mole in I_3 between manganese and iron, $E^0[Fe^{3+}/Fe^{2+}]$ is as high as 0.77 V. Consequently, at the next element, where $E^0[Co^{3+}/Co^{2+}] = 1.95$ V, the renewed increase in ΔH^0 for the $M^{II} \rightarrow M^{III}$ process is sufficient to ensure that the dipositive ion is more stable than Mn^{2+} with respect to the tripositive state. By contrast, the value of $E^0[M^{3+}/M^{2+}]$ is higher for europium than for any other lanthanide metal. Even at ytterbium, after a gap of seven elements, the stability of dipositive europium has not been exceeded. Owing to the absence of large ligand field stabilization energies, the value of ΔH^0 for the $M^{II} \rightarrow M^{III}$ process feels nearly the full impact of the drop in I_3 after the half-filled shell, and this substantial decrease cannot be compensated by the relatively mild increases in ΔH^0 from gadolinium onwards.

6.9. The oxidation states of the actinides. Some electrode potentials relevant to a discussion of the oxidation states of the actinides in aqueous solution are given in fig. 6.10. Formation of the oxycations MO_2^+ and MO_2^{2+} may be mentally visualized as initial production of the gaseous M^{5+} and M^{6+} ions, followed by the formation of chemical bonds with oxide anions and subsequent hydration. Provided that the first step does not require removal of an electron from the stable radon core, then the others more than compensate for it and, early in the series, these complexes strongly resist reduction. However, with increasing atomic number, they steadily become less stable with respect to the tetrapositive and tripositive ions while beyond americium they are unknown. Consequently, the M^{4+}/M^{3+} electrode potential is the quantity upon which the oxidation–reduction behaviour of the actinides turns.

$$PaO_2^+ \xrightarrow{\ -0.1\ } Pa^{4+}$$

$$UO_2^{2+} \xrightarrow{\ 0.06\ } UO_2^+ \xrightarrow{\ 0.58\ } U^{4+} \xrightarrow{\ -0.63\ } U^{3+}$$

$$NpO_2^{2+} \xrightarrow{\ 1.14\ } NpO_2^+ \xrightarrow{\ 0.74\ } Np^{4+} \xrightarrow{\ 0.16\ } Np^{3+}$$

$$PuO_2^{2+} \xrightarrow{\ 0.91\ } PuO_2^+ \xrightarrow{\ 1.17\ } Pu^{4+} \xrightarrow{\ 0.98\ } Pu^{3+}$$

$$AmO_2^{2+} \xrightarrow{\ 1.60\ } AmO_2^+ \xrightarrow{\ (1)\ } Am^{4+} \xrightarrow{\ (>2)\ } Am^{3+}$$

$$Bk^{4+} \xrightarrow{\ 1.6\ } Bk^{3+}$$

Fig. 6.10. The electrode potentials of the actinide ions.

It is evident that the variations in this potential may be interpreted with the argument symbolized by fig. 6.9 if the lanthanides on the horizontal axis are replaced by their actinide analogues, and the graphs are shifted one element to the right. The ligand field stabilization energies are two to five times greater than in the lanthanide series, but they are still small, and as those ionic radii which are known diminish steadily across the series, a smooth variation in the differences in the hydration enthalpies is a reasonable assumption.

Thorium tri-iodide has been prepared by heating the metal with the tetra-iodide but it reduces water vigorously, while although the same reaction occurs in solutions containing U^{3+}, it does so with less evolution of heat. Tripositive neptunium is stable indefinitely in de-aerated water but it is completely destroyed when oxygen is admitted. Pu^{3+} is unaffected under these conditions and requires quite powerful oxidizing agents like bromate and permanganate to convert it to the tetrapositive state. The aquated ions Am^{4+} and Cm^{4+} are unknown and the tetrafluorides oxidize water unless it contains a high concentration of fluoride ions. Under these circumstances, fluoride complexes of americium(IV) and curium(IV) are thought to exist, but although the first is stable, the second slowly oxidizes the solvent. If it is assumed that the overall stability constants of the two complexes are not very different, then because the oxygen overvoltage is exceeded only with the curium compound it is reasonable to suppose that the M^{4+}/M^{3+} potential is greater for this element. From curium to berkelium the drop in the fourth ionization enthalpy is apparent in the value of 1·6 V for the Bk^{4+}/Bk^{3+} potential, but from californium to mendelevium, there is no evidence for the existence of any +4 oxidation state.

All these observations and the known electrode potentials are consistent with the following stability sequence for the tetra-positive ions:

Th > Pa > U > Np > Pu > Am > Cm < Bk > Cf etc.

The Racah parameters, a measure of the interactions between the electrons can be determined spectroscopically for some of these ions. The limited information which is available at present suggests that they are lower than in the lanthanide series. This is consistent with an apparently smaller drop in the fourth ionization potential after the half-filled shell which, moving back through the series, causes the stability of berkelium(IV) compounds to be surpassed at plutonium(IV). With the lanthanides, only cerium(IV) exceeds terbium(IV) in its stability with respect to reduction to the tri-positive state. The possibility of a substantial difference in the general overall increases in the enthalpies of corresponding M^{III}–M^{IV} processes in the lanthanide and actinide series has been ignored in this last argument (compare page 141).[1]

The speculative discussions in §6.8 and §6.9 assume that terms such as $(\Delta H_h^0[M^{3+}] - \Delta H_h^0[M^{4+}])$ and $(U[MX_3] - U[MX_2])$ vary smoothly, or at least less violently than other equally speculative terms which involve ionization potentials. Such assumptions seem reasonable, but only accurate thermodynamic data, at present unavailable, can show whether they are justified.[2]

[1] Recent preparation of aqueous Md^{2+} and very stable No^{2+} suggests that the argument implied by Fig. 6.9 can be extended to depositive actimides.

[2] The principles outlined in §6.8 and **6.9** were developed in collaboration with Dr P. G. Nelson.

6.10. Cyanide complexes of the first transition series and the hexafluoro-metallates(IV). The problems that were discussed in §6.6–6.9 showed the importance of the appropriate ionization potential in determining the variations in the stabilities of two oxidation states across a horizontal period. Unfortunately this importance cannot be promoted to the status of a general principle. This may be illustrated by a problem involving some cyanide complexes of the metals of the first transition series.

For chromium, manganese, iron and cobalt, the potentials of the M^{III}-cyanide/M^{II}-cyanide couples which, apart from cobalt(II), are thought to contain hexacyano-complexes, are $-1\cdot3$, $-0\cdot22$, $+0\cdot36$ and $< -0\cdot8$ volts respectively. The $Cr(CN)_6^{4-}$ ion liberates hydrogen from water in the presence of a platinum catalyst, and the manganese(II) and cobalt(II) cyanide complexes, in contrast to the stability of ferrocyanide, are oxidized in solution by air. The stability sequence for the $+2$ state, $Cr < Mn < Fe > Co$, therefore differs considerably from that in the third ionization potential ($Cr < Mn > Fe < Co$). When the problem is restated by a cycle similar to that in fig. 6.6, very large ligand field stabilization energies, magnified by spin-pairing (see §6.11), are believed to be the cause of the increased influence of the metal–ligand interaction terms (Chadwick & Sharpe, 1966). The d^6 configuration goes low-spin at comparative small ligand field strengths (see page 147) and thereafter its crystal field stabilization energy increases as $\frac{12}{5}\Delta$. Although very large values of Δ in the $+3$ oxidation state tend to lower the potentials of the M^{III}-cyanide/M^{II}-cyanide couples, this stabilization of the $M(CN)_6^{3-}$ ion is less in the case of iron because $Fe(CN)_6^{4-}$ is a d^6 configuration in a strong ligand field. When water is replaced by cyanide in an M^{III}/M^{II} couple, many energetic factors are involved, but it is mainly these ligand field effects which are believed to be responsible for the stability sequence $Mn < Fe > Co$ for the $M(CN)_6^{4-}$ ions, and which, in particular, more than compensate for the drop in the third ionization potential between manganese and iron.

Another problem of interest is that of the stabilities of the compounds A_2MF_6, where A is an alkali metal and M is a member of the first transition series. This appears to contain features which make it intermediate in character between the problems of the variations in the potentials of the M^{3+}/M^{2+} and M^{III}-cyanide/M^{II}-cyanide couples. The reader is warned that the discussion that follows is necessarily speculative, mainly because there is an almost complete absence of thermochemical data for these compounds, and because their modes of decomposition are not well established. Let us suppose that they are unstable to processes of the type:

$$A_2MF_6(s) \rightarrow A_2MF_5(s) + \tfrac{1}{2}F_2(g). \qquad\qquad [6.15]$$

Some guesses at the variation in ΔG^0 may be made from the chemistry of the compounds. K_2TiF_6 is very stable to heat, the corresponding vanadium compound is made by fluorination of K_2VF_5 at 500°, K_2CrF_6 decomposes at 300°, a temperature below that at which K_2MnF_6 can be prepared, while no fluoride complexes of iron(IV) are known. Among the compounds M_2CoF_6, only the caesium salt, the most stable on an ionic model, has been prepared, while the potassium, rubidium and caesium analogues are known for nickel. No compounds of this type are known for copper, zinc or gallium. These observations are consistent with the stability order:

Ti > V > Cr < Mn > Fe < Co < Ni > Cu, Zn, Ga.

We shall assume that the variations in the stabilities of the A_2MF_6 compounds along the series are dominated principally by the changes in the fourth ionization potential, and that this increases fairly smoothly from titanium to iron, drops between iron and cobalt after the half-filled shell, and then renews its steady increase from cobalt to gallium. This implies the stability sequence Ti > V > Cr > Mn > Fe < Co > Ni > Cu, Zn, Ga which clashes with our tentative experimental order only between chromium and manganese and between cobalt and nickel.

The first discrepancy corresponds to the vanadium-chromium step in the problem of §6.6, and is the point where the crystal field stabilization energies of high-spin compounds exert their maximum effect. In addition K_2NiF_6 is a low-spin d^6 complex with a very large ligand field stabilization energy. The crystal field splittings have a greater effect in this system than in those of §6.6 and §6.7 because the oxidation number of the oxidized and reduced states has increased by one, so the absolute values of the splittings, and the differences between them for two oxidation states of the same metal are considerably larger. Thus the stability of the A_2MF_6

compounds with respect to the tripositive oxidation state can be treated by extending the conclusions of the problem in §6.6, and modifying them slightly to allow for the larger ligand field effects.

6.11. Spin-pairing in transition metal complexes. The problems covered in the last seven sections mainly involved high-spin complexes. The alternative low-spin electronic arrangement which is possible for the d^4, d^5, d^6 and d^7 configurations in O_h symmetry, was only briefly mentioned in §6.3 and §6.10.

When an octahedral transition metal complex passes into a low-spin state, the number of electrons in the d shell is unaltered but the spin is partially or wholly quenched. Thus from the arguments in §6.6, although the coulombic repulsion term remains roughly constant, there is an unfavourable decrease in the exchange energy which must be exceeded by the orbital energy gained through the movement of electrons from the e_g to the t_{2g} levels. If the number of electrons involved in this process is n, then the condition for spin pairing is,

$$n\Delta > K\delta m, \tag{6.10}$$

where K and δm have the meanings of §6.6. Forgetting other approximations that will be mentioned later, this assumes that the major sources of the binding energy, such as the coulombic attraction and repulsion between the ligands and the central ion, are unaffected by the change of spin state. According to molecular orbital theory, electrons are transferred from anti-bonding e_g to t_{2g} orbitals of lower energy when a complex goes low-spin. This should lead to a shortening of the metal–ligand bond lengths, and there is strong evidence that this occurs. Thus caesium cobalt(III) alum, which contains the only low-spin tripositive aquo-complex in the first transition series, has a cell side that is noticeably smaller than those of all the other isomorphous, cubic, caesium alums of this series. Such changes must affect energy terms other than those considered in (6.10), but nevertheless the equation is still accurate enough to yield one interesting result.

For the configurations d^4, d^5, d^6 and d^7 in an octahedral field, δm takes the values three, six, four and two respectively, while n is one for d^4 or d^7 and two for d^5 or d^6. Thus, as the value of Δ is increased, d^6 and d^7 should pair their spins before d^4 and d^5. This crude approach works well in a comparison of the d^5 and d^6 cases, as may be demonstrated by considering the points in the spectrochemical series where ions with these configurations and the same charge pass over into low-spin behaviour. Thus while manganese(II) pairs its spins between nitrite and cyanide, iron(II) does so between ethylenediamine and phenanthroline. Again CoF_6^{3-} is the only known high-spin cobalt(III) complex, but a ligand with a field strength between ethylenediamine and phenanthroline is required for the formation of low-spin iron(III).

Once a d^6 system has turned low-spin, its orbital stabilization increases as $\frac{12}{5}\Delta$. This rapid increase, combined with the tendency to pair spins at weak field strengths, will often raise the stability constants of low-spin d^6 species relative to those of the analogous complexes of other transition metals. This is the source of the anomalous position of $Fe(phen)_3^{2+}$ in fig. 6.5 and is undoubtedly the major contributory factor to the high stabilities of cobalt(III) complexes with strong field ligands. The overall stability constants of $Fe(EDTA)^-$ and $Co(EDTA)^-$ are 10^{25} and 10^{36} mole^{-1} litre respectively, and those of $Fe(phen)_3^{3+}$ and $Co(phen)_3^{3+}$ are 10^{15} and 10^{45} mole^{-3} litre^3.

It would be interesting to treat the variations in the relative stabilities of two oxidation states of the same metal as the ligand is altered by means of table 6.2, the spectrochemical series, and (6.10). Ignoring the fact that K, which can be expressed in terms of the Racah parameters B and C, is not a true constant but varies with the ligands, such a treatment neglects the changes in the large proportion of the binding energy that is not due to ligand field effects. These large approximations make it difficult to give a reliable treatment of the subtle variations in the relative stabilities of oxidation states as the ligands alter and the spin states change. Thus $E^0[Fe^{3+}/Fe^{2+}] = 0.77$ V, but with phenanthroline and cyanide, two ligands of comparable and much higher field strengths, the standard potentials of the couples $Fe(CN)_6^{3-}/Fe(CN)_6^{4-}$ and $Fe(phen)_3^{3+}/Fe(phen)_3^{2+}$ are 0.36 and 1.06 V and lie on either side of the figure for the aquated ion system. The difference between these two potentials is equivalent to only about 14 kcal/mole, and it would obviously be very difficult to pick out this relatively small quantity from the several, individually large thermodynamic factors that are involved.

6.12. The relative stabilities of oxidation states down a transition metal group. The variations in the relative stabilities of two oxidation states of the same metal down a vertical transition metal group is a very general problem. As the ligands, the oxidation states and the group are changed, almost every possible pattern of behaviour is discernible in moving from the first, through the second and into the third transition series. We may first quote examples in which the relative stability of the higher of two oxidation states increases, the ligand remaining unchanged. Many experimental examples are known. Thus potassium permanganate loses oxygen at 200° yielding MnO_2 while the corresponding pertechnetate and perrhenate sublime unchanged at over 1000°; all the hexahalides of tungsten have been prepared, but only the hexafluoride and hexachloride of molybdenum, and the hexafluoride of chromium are known; the highest fluorides of nickel, palladium and platinum are the compounds NiF_2, PdF_4 and PtF_6; Fe_3O_4, RuO_2 and OsO_4 are the oxides formed when the metals are heated in dry air.

On the other hand, exceptions to this pattern are also known. While $E^0[M^{3+}/M^{2+}]$ is 0·77 V for iron and 0·25 V for ruthenium, the values for the $M(CN)_6^{3-}/M(CN)_6^{4-}$ couples are 0·36 and 0·9 V respectively. The replacement of water by cyanide reverses the order of potentials, and the higher oxidation state then becomes less stable from iron to ruthenium.

If the increasing stability of higher oxidation states down a group were a general principle, then low or intermediate oxidation states should at the same time become less stable with respect to disproportionation. Thus if manganese dioxide is heated in vacuo, it decomposes to MnO, but rhenium dioxide yields the metal and the heptoxide; the non-existence of platinum difluoride is probably due to its ready disproportionation into the metal and the tetrafluoride, but the nickel and palladium analogues are known. Exceptions to this behaviour may also be found. While no monohalides of iron are known, the monoiodides of ruthenium and osmium have recently been prepared. Their low magnetic moments suggest that metal–metal bonds are present. Irregular stability variations also occur; compounds of copper(I) and gold(I) disproportionate much more readily into the metal and higher oxidation states than do those of silver(I).

This variety of observed stability patterns reflects the very general nature of the question and its consequent difficulty. The problem of the variations in the relative stabilities of two different oxidation states of the same metal down a vertical transition metal group may be restated by means of a cycle similar to that expressed by (6.5). An enthalpy change, ΔH, observed when a compound or complex of a metal in a lower oxidation state, a, is converted into another of the same metal in a higher oxidation state, b, may be viewed as the sum of three terms. If ΔH_b is the enthalpy change associated with the formation of the gaseous metal ion, M^{b+}, the gaseous ligands and any residual gaseous species from the complex containing the higher oxidation state, ΔH_a is the corresponding term for the reduced complex and ΔH_i is the enthalpy of the process:

$$M^{a+}(g) \rightarrow M^{b+}(g) + (b-a)e^-(g), \qquad [6.16]$$

then for a reaction at constant temperature, constant a and constant b with a particular oxidizing system

$$\Delta G = -(\Delta H_b - \Delta H_a) + \Delta H_i - T\Delta S + \text{constant}. \qquad (6.11)$$

The variety of behaviour encountered earlier in this section implies that the variations in either $(\Delta H_b - \Delta H_a)$ or ΔH_i may be of over-riding importance as one moves down a vertical transition metal group. However if any salient pattern does emerge, it is the increasing stability of the very highest oxidation states. This may be correlated with the tendency of the higher ionization potentials to fall in moving from the first, through the second and into the third transition series. Apart perhaps from this, reliable generalizations cannot be made. Furthermore, an interpretation of the changes in either the $(\Delta H_b - \Delta H_a)$ or ΔH_i terms with reliable predictive powers does not yet exist. For these reasons, the problem will not be considered further.

6.13. The oxidation states of transition elements and typical metals. One common property of transition elements is their willingness to form complexes with the same ligand in a number of different oxidation states. If equation (6.11) is used to compare the abilities of a transition element and a typical metal in this respect, the ionization potential is very often a factor of major importance. In fig. 6.11, successive ionization potentials of vanadium, gallium and arsenic are plotted against the charge on the ion that is formed. Values of $(\Delta H_b - \Delta H_a)$ will obviously vary substantially from ligand to ligand, and with the physical state of

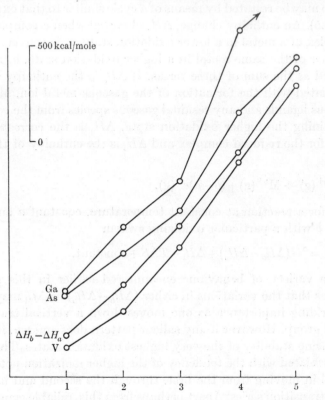

Fig. 6.11. The ionization potentials of arsenic, gallium and vanadium with an example of a possible variation in $(\Delta H_b - \Delta H_a)$ (see text).

the complex or compound concerned. A rough, general idea of the way in which it changes with n in some cases can probably be gained in the following way. From the heats of formation of the compounds NaF, MgF_2, AlF_3, SiF_4 and PF_5, the standard enthalpies of the process,

$$RF_n \rightarrow R^{n+}(g) + nF^-(g) \tag{6.17}$$

may be calculated via the Born–Haber cycle. The values of $\Delta H^0[RF_n] - \Delta H^0[RF_{n-1}]$ are plotted against n in fig. 6.11 and, for the purposes of this section, may be assumed to provide some rough indication of the type of variation in $(\Delta H_b - \Delta H_a)$ that might be expected when $b = a + 1$. To facilitate comparisons, all the plots have different zero lines.

Both the vanadium and $(\Delta H_b - \Delta H_a)$ plots approach curves of smoothly increasing gradient, the differences between them remaining quite similar. Under these circumstances, the stabilities of successive oxidation states with respect to the ones preceding them will not differ immensely. This is consistent with the capacity of a transition metal to form complexes with the same ligand in a number of different oxidation states. Thus vanadium forms the fluorides VF_2, VF_3, VF_4 and VF_5, and chromium yields all fluorides from oxidation state two to oxidation state six. Dipyridyl complexes of the formulae $Cr(dipy)_3$, $Cr(dipy)_3^+$, $Cr(dipy)_3^{2+}$ and $Cr(dipy)_3^{3+}$ have also been characterized.

In contrast, the gallium and arsenic plots show marked discontinuities which are apparently associated with the stabilities of filled p and s shells. These irregularities help to make all oxidation states except one and three unstable to these two, and rather exceptional variations in the $(\Delta H_b - \Delta H_a)$ term would be required to offset them. For gallium in fact, the oxidation state $+3$ is the only one with any extensive existence, while the chemistry of arsenic is almost entirely that of oxidation states $+3$ and $+5$.

6.14. Summary. In the main part of this chapter, some problems concerned with the relative stabilities of certain oxidation states have been restated by means of (6.11). In §**6.6–6.9**, this enabled several apparently unconnected phenomena to be reduced to one—a saw-tooth variation in the appropriate ionization potential. This correlation is a very valuable one, although the theoretical interpretations of the changes in ΔH_i and $(\Delta H_b - \Delta H_a)$ were only qualitative.

In §**6.10**, two problems which were at first sight rather similar to those in §**6.6–6.9** were discussed. Here one portion of the total metal–ligand interactions, the ligand field stabilization term, appeared to reduce, or at points to almost eliminate the influence of ΔH_i on the order of relative stabilities.

The restatement was again fairly illuminating in §**6.13** for the variations in ΔH_i were of great importance, but was almost worthless in §6.12 where the competing influences of the $(\Delta H_b - \Delta H_a)$ and ΔH_l terms were much more equal and the problem was very general. In all cases, the absence of adequate theoretical interpretations of the variations in $(\Delta H_b - \Delta H_a)$ or ΔH_l and the lack of important thermodynamic data were a severe handicap.

References and suggestions for further reading

Ballhausen, C. J. (1962). *Introduction to Ligand Field Theory*. New York: McGraw-Hill.

Chadwick, B. M. & Sharpe, A. G. (1966). *Advances in Inorganic Chemistry and Radiochemistry*. Editors: Emeleus, H. J. & Sharpe, A. G., vol. 8, p. 108. New York: Academic Press.

Cotton, F. A. (1964). *J. Chem. Educ.* **41**, 466. An excellent elementary introduction to ligand field theory.

Habermann, C. E. & Daane, A. H. (1964). *J. Chem. Phys.* **41**, 2818.

Jorgensen, C. K. (1962a). *Absorption Spectra and Chemical Bonding in Complexes*. Oxford: Pergamon Press.

Jorgensen, C. K. (1962b). *Mol. Phys.* **5**, 271.

Jorgensen, C. K. (1964). *Mol. Phys.* **7**, 417.

Nancollas, G. H. (1966). *Interactions in Electrolyte Solutions*. London: Elsevier.

Nelson, P. G. & Sharpe, A. G. (1966). *J. Chem. Soc.* (A), 501. The paper on which the discussions in §6.6–6.10 are modelled.

Orgel, L. E. (1966). *An Introduction to Transition Metal Chemistry; Ligand Field Theory*. 2nd edition. London: Methuen. A non-mathematical account of ligand field theory at the undergraduate level.

Stability Constants (1964). Chem. Soc. Special Publication No. 17, London.

Watson, R. E. & Freeman, A. J. (1964). *Phys. Rev.* **134**A, 1526.

7. Bond Energies and some Aspects of Non-Metal Chemistry

7.1. The chemical bond. The origins of the concept of the chemical bond preceded the introduction of wave mechanics by more than fifty years. Shortly after 1850, Frankland noted that the elements combined with characteristic numbers of equivalents of other atoms, and introduced the term 'bond' as a numerical unit of this quantitative combining power. Couper and Kekulé independently showed how this idea systematized the then clumsy structure of organic chemistry, while the former initiated modern structural formulae by introducing the line between two atoms as a graphic description of a single bond. At this stage, diagrams of this kind were regarded as no more than a convenient expression of the stoicheiometry of a molecule and the observed chemical equivalence or non-equivalence of the atoms contained within it. In 1874, by drawing heavily on the investigations of optical activity made by Pasteur, van't Hoff and Le Bel concluded that the four hydrogen atoms in methane lay at the corners of a regular tetrahedron. This recognition of the three-dimensional distribution of valencies, which was extended to include transition metal complexes by Werner, gave Couper's formulae a spatial reality that they had previously lacked.

If was some fifty years before the advent of X-ray crystallography and electron diffraction confirmed directly the results that van't Hoff, Le Bel and Werner had deduced from their studies of stereoisomerism. They viewed chemical compounds as assemblies of atoms linked together, according to certain rules, by chemical bonds of a numerical order. In the field of non-metallic chemistry, this idea has been a very helpful one. Its power has been greatly extended by describing the structures of compounds in terms of models based upon the electronic structures of atoms. These models originated in 1916 with the twin publications of Kossel and Lewis.

TABLE 7.1 *S–S distances in some sulphur compounds*

Compound	S–S distance (Å)	Compound	S–S distance (Å)
S_2H_2	2·05	$Na_2S_4O_6.2H_2O$	2·02
S_2Cl_2	1·98	$C_2H_6S_2$	2·04
S_2Br_2	1·97	$C_2H_6S_3$	2·04
S_8	2·07	$C_2F_6S_2$	2·05
$BaS_4O_6.2H_2O$	2·02	$C_2F_6S_3$	2·07
$BaS_5O_6.2H_2O$	2·04	—	—

The work of Lewis was of greater relevance to the chemistry of the non-metals, and he placed particular emphasis on the description of single bonds as pairs of electrons shared between the two atoms.

The relation of chemical bonds to electron pairs suggested that the bonds themselves might have a certain individuality. In graphic representations of the structures of molecules, the most significant characteristics of a chemical bond are its multiplicity and the two atoms at either end of it, so the next logical step was to see whether its properties depended only upon these two characteristics. This would imply, for example, that the C—C single bond in C_2Cl_6 does not differ in any discernible way from that in ethyl chloride or propylene. At that time, chemists were interested mainly in the nuclear separation of the two bound atoms and some measure of the force holding these atoms together. A definition of the first property presents no fundamental difficulty, and the distances between atoms can be determined fairly precisely by the techniques of X-ray crystallography, electron diffraction and microwave spectroscopy. The results that were obtained, such as the S—S bond distances for divalent sulphur in table 7.1, suggested that the assumption that bond length is independent of secondary environment is not precisely valid, but that it is sufficiently correct to be useful. It is much more difficult to obtain a measure of bond strength. Some quantities which have been suggested for this purpose are discussed in the next four sections.

7.2. Bond dissociation energies. The bond dissociation energy is defined here as the enthalpy change associated with the reaction in which one mole of the bond is homolytically broken, reactants and products being in the ideal gas state at one atmosphere pressure and 25°. Thus for methane, the C—H bond dissociation energy, $D(CH_3—H)$, is 104 kcal/mole:

$$CH_4(g) \rightarrow CH_3^{\cdot}(g) + H^{\cdot}(g), \quad \Delta H^0 = +104 \text{ kcal/mole}. \qquad [7.1]$$

This quantity may be determined directly by one of a number of experimental techniques. In [7.1] however, the methyl radical is almost certainly planar, while theoretical interest resides mainly in the reaction where it retains the same shape as the —CH$_3$ group in methane. Thus the bond dissociation energy includes the energies of inconvenient stereochemical and electronic rearrangements which rather reduces its use as a thermochemical measure of bond strength. It is perhaps partly for this reason that, except in favourable circumstances, bond dissociation energies show a considerable dependence on the secondary environment. Compared with the methane figure given above, the values for the successive loss of a hydrogen atom from the radicals $CH_3^{\cdot}$, $CH_2^{\cdot}$ and $CH^{\cdot}$ are 106, 106 and 81 kcal/mole. Again, for H_2O and $OH^{\cdot}$, the values for the O—H bond are 118 and 103 kcal/mole respectively.

In more complex carbon chemistry, the uniform stereochemistry sometimes gives rise to a great similarity in the immediate secondary environment of bonds, and a remarkable constancy is often observed in bond dissociation energies. For example, in paraffin hydrocarbons RCH_3 where $R \neq H$, the bond dissociation energies $D(RCH_2—H)$ all lie within 1 kcal/mole of 98 kcal/mole. Empirically based confidence in the reproducibilities of these and other bond dissociation energies has enabled the theory of the structures of organic compounds to reach a highly sophisticated stage. In propylene, for example, the value of $D(CH_2CHCH_2—H)$ is only 85 kcal/mole and the lowering of 13 kcal/mole is attributed to the resonance energy of the allyl radical $CH_2{=}CH—CH_2^{\cdot}$ (see Benson, 1965).

7.3. Bond energy terms. Bond energy terms are quantities assigned to each of the bonds in a molecule such that the sum over all bonds is equal to the enthalpy change associated with the conversion of the molecule into separate atoms. Reactants and products are in their ideal gas states at one atmosphere pressure and

25°. If individual values are required, this definition is inadequate and the assignment is usually made by introducing the additivity assumption, namely that the energy term for a given bond is constant from molecule to molecule.

The bond energy term is only equal to the bond dissociation energy for diatomic molecules. Thus for methane,

$$CH_4(g) \rightarrow C(g) + 4H(g), \quad \Delta H^0 = 397 \cdot 6 \text{ kcal/mole} \qquad [7.2]$$

and as there are four C–H bonds in the molecule the bond energy term, $B(C—H)$ is $99 \cdot 4$ kcal/mole. This could be combined with $\Delta H_f^0[C_2H_6(g)]$ to yield $B(C—C)$:

$$C_2H_6(g) \rightarrow 2C(g) + 6H(g). \qquad [7.3]$$

$$\Delta H^0 = 2\Delta H_f^0[C(g)] + 6\Delta H_f^0[H(g)] - \Delta H_f^0[C_2H_6(g)] \qquad (7.1)$$

$$= 675 \cdot 4, \qquad (7.2)$$

$$= 6B(C—H) + B(C—C), \qquad (7.3)$$

$$= 596 \cdot 4 + B(C—C), \qquad (7.4)$$

$$\therefore B(C—C) = 79 \cdot 0 \text{ kcal/mole.} \qquad (7.5)$$

If, in accordance with the additivity assumption, bond energy terms were dependent only upon the multiplicity of the bond and the atoms at either end of it, then the two values of $B(C—H)$ and $B(C—C)$, together with the heats of atomization of carbon and hydrogen, would reproduce, within the limits set by experimental uncertainties, the standard heats of formation of every known paraffin hydrocarbon in the gas phase.

Unfortunately, mean bond energy terms are not additive to that extent, and this immediately introduces an element of imprecision into the definition given at the opening of this section. From the two values given above, the enthalpy of atomization of neo-pentane is $4B(C—C) + 12B(C—H)$ or $1508 \cdot 8$ kcal/mole. As

$$5\Delta H_f^0[C(g)] + 12\Delta H_f^0[H(g)]$$

is $1481 \cdot 7$ kcal/mole, we obtain $-27 \cdot 1$ kcal/mole for

$$\Delta H_f^0[\text{neo-}C_5H_{12}(g)]$$

compared with an experimental value of $-39 \cdot 7$ kcal/mole. Thus we are faced with an immediate problem: should we calculate

$B(C—C)$ from neo-pentane, accept the value derived from ethane or adjust $B(C—H)$ so that $B(C—C)$ is the same in both compounds? This specific question, then, is a reflection of a weakness that is implicit in the concept of a bond energy term. Unambiguous values can only be obtained if the additivity assumption is completely justified by experiment. As the additivity is not precise, the values of bond energy terms are dependent upon the order or system by which they are derived.

Variations in bond energy terms and the consequent departures from additivity are particularly severe when an atom at the end of a bond changes its valency. In the reaction:

$$Cl_2(g) + ClF_3(g) \rightarrow 3ClF(g), \qquad [7.4]$$

the number of Cl—F bonds remains constant while one Cl—Cl bond is destroyed. If the bond energy terms were the same in ClF and ClF_3, then the standard enthalpy of this reaction would be $+58\cdot2$ kcal/mole, the dissociation energy of chlorine. Owing to the increase in the number of moles of gas, the reaction has a favourable entropy change of $35\cdot6$ cal/deg.mole but this would be wholly insufficient to yield a negative free energy change. In practice, [7.4] forms the basis of an excellent method for the preparation of chlorine monofluoride at 250°. The bond energy terms calculated from $\Delta H_f^0[ClF(g)]$ and $\Delta H_f^0[ClF_3(g)]$ are 60 and 42 kcal/mole so that the higher fluoride is much less stable with respect to the lower than a unique Cl—F bond energy term would imply.

Nevertheless, within a range of compounds formed by atoms in fixed valencies, the assumption of additivity in mean bond energies appears to be a useful one. Tetravalent carbon in particular forms a very large number of compounds containing very few different bonds. Thus the additivity approximation allows a small number of bond energy terms to reproduce fairly closely a very large number of heats of formation. Consequently, in organic chemistry, considerable trouble has been taken to derive values for bond energy terms which minimize the departures from additivity over a considerable range of compounds. As with bond dissociation energies, confidence in the constancy of mean bond energies has reached a stage where notable departures are often attributed to resonance effects. From the bond energy terms in table 7.2, the

TABLE 7.2 *Bond energy terms*

Bond	Value (kcal/mole)[a]		Compound(s)[g]
	Experimental	Calculated[b]	
$O{=}O$	119·1	—	O_2
H—H	104·2	—	H_2
O—H	111	108	H_2O
O—O	34	—	H_2O_2
F—F	37·8	—	F_2
F—O	51	38	F_2O
F—H	135·8	138	HF
Cl—Cl	58·2	—	Cl_2
Cl—O	49	49	Cl_2O
Cl—H	103·3	103	HCl
Cl—F	60	60	ClF
Br—Br	46·1	—	Br_2
Br—H	87·5	86	HBr
Br—F	60	62	BrF
I—I	36·2	—	I_2
I—H	71·3	70	HI
I—F	58	70	IF
I—Cl	50	51	ICl
I—Br	42	42	IBr
$S{=}S$	102·6	—	S_2
S=O	125	—	SO
S—H	88	88	H_2S
S—S	63	—	S_8, various H_2S_n
S—Cl	65	66	SCl_2
Se—Se	38	—	Se_2Cl_2, Se_2Br_2
Se—H	73	66	H_2Se
Se—Cl	58	58	$SeCl_2$
$N{\equiv}N$	226·0	—	N_2
$N{\equiv}O$	151·0	—	NO
$N{\equiv}O^+$	251·7[c]	—	NO^+
N—H	93	85	NH_3
N—O	39[d]	42	NH_2OH
N—F	67	55	NF_3
N—Cl	45[d]	47	NCl_3
N—N	38	—	N_2H_4, N_2F_4
N=O	142	—	NOF, NOCl
$P{\equiv}P$	125	—	P_2
P—H	78	72	PH_3
P—P	50	—	P_4, P_2H_4
P—F	119	118	PF_3
P—Cl	79	79	PCl_3
P—Br	64	64	PBr_3
P—O	88[e]	89	P_4O_6
As—As	43	—	As_4
As—O	79	86	As_4O_6
As—F	116	115	AsF_3

For notes see p. 160.

Tᴀʙʟᴇ 7.2 (*cont.*)

Bond	Value (kcal/mole)[a]		Compound(s)[g]
	Experimental	Calculated[b]	
As—Cl	74	75	$AsCl_3$
As—Br	61	61	$AsBr_3$
As—H	71	67	AsH_3
Sb—Sb	34	—	Sb_4
Sb—Cl	75	74	$SbCl_3$
Sb—Br	63	61	$SbBr_3$
Sb—H	61	60	SbH_3
C—C	82·9	—	Organic[f]
C=C	146	—	Organic[f]
C≡C	200	—	Organic[f]
C—H	98·8	96	Organic[f]
C—N	72·9	67	Organic[f]
C—O	85·6	87	Organic[f]
C=O	176	—	Aldehydes
C—O	179	—	Ketones
C=O	192	—	CO_2
C=O	257·3	—	CO
C—F	117	111	CF_4
C—Cl	78	84	CCl_4
C—Br	65	70	CBr_4
Si—H	76⎫	76	⎧$Si(c), SiH_4$
Si—Si	54⎭		⎩Si_2H_6, Si_3H_8
Si—O	111	109	$SiO_2(c)$
Si—O	153		SiO_2
Si=O	192	—	SiO
Si—F	143	142	SiF_4
Si—Cl	96	95	$SiCl_4$
Si—Br	79	78	$SiBr_4$
Si—C	73	73	$Si(CH_3)_4, SiC(s)$
Si—N	80	79	$Si_3N_4(s)$
Ge—Ge	45⎫	68	⎧$Ge(s), GeH_4$
Ge—H	68⎭		⎩Ge_2H_6, Ge_3H_8
Ge—O	86	87	$GeO_2(s)$
Ge—F	113	115	GeF_4
Ge—Cl	81	77	$GeCl_4$
Ge—Br	67	62	$GeBr_4$
Ge—I	51	48	GeI_4
Ge—N	61	62	$Ge_3N_4(s)$
Sn—Sn	36	—	$Sn(s)$
Sn—H	60	61	SnH_4
Sn—C	50	58	$Sn(CH_3)_4, Sn(CH_3)_3H$
Sn—Cl	75	76	$SnCl_4$
Sn—Br	64	62	$SnBr_4$
B—F	154	154	BF_3
B—Cl	106	106	BCl_3
B—Br	88	87	BBr_3

Table 7.2   (*cont.*)

| Bond | Value (kcal/mole)[a] | | Compounds[g] |
	Experimental	Calculated[b]	
B—I	65	65	BI_3
B—B	72	—	B_2F_4
B—O	125	120	Various boron esters, ester chlorides and hydroxides

[a] Bond energy terms for diatomic molecules, which are equal to the bond dissociation energies, are given to one place of decimals where the uncertainty is of this order.
[b] Calculated from (7.10) using the figures in table 7.6.
[c] For dissociation into N and O^+.
[d] Using the heats of vaporization estimated by Cottrell (1958).
[e] Using the heat of vaporization quoted by S. B. Hartley *et al.* (1963).
[f] Values obtained from organic compounds were taken from Coates and Sutton (1948) and adjusted to the heats of atomization in table 7.3.
[g] Compounds quoted are in the gas phase unless otherwise stated.

standard enthalpy change associated with the atomization of the hypothetical gaseous molecule of cyclohexatriene is given by

$$\Delta H^0 = 3B(\text{C}=\text{C}) + 3B(\text{C}-\text{C}) + 6B(\text{C}-\text{H}) \tag{7.6}$$

$$= 1280 \text{ kcal/mole.} \tag{7.7}$$

For benzene, $\Delta H_f^0[C_6H_6(g)] = +19\cdot8$ kcal/mole, and with the known values of $\Delta H_f^0[C(g)]$ and $\Delta H_f^0[H(g)]$, the enthalpy of atomization is

$$\Delta H^0 = 6 \times 171\cdot3 + 6 \times 52\cdot1 - 19\cdot8 \tag{7.8}$$

$$= 1321 \text{ kcal/mole.} \tag{7.9}$$

The difference of 41 kcal/mole between (7.7) and (7.9) is usually credited to the resonance energy of benzene.

In the inorganic chemistry of the non-metals, the careful assignment of bond energy terms has been seen as a less profitable task because the elements often display different valencies in their compounds, and because in any particular valency, by comparison with carbon, relatively few compounds are formed with other nonmetals. Thus a substantial condensation of thermochemical infor-

TABLE 7.3 *Standard enthalpies of atomization of some elements* (*kcal/mole*)

Element	ΔH_f^0 [X(g)]	Element	ΔH_f^0 [X(g)]
As	72·3	N	113·0
B	134·5	O	59·6
Br	26·7	P	79·8[a]
C	171·3	S	66·6
Cl	29·1	Sb	62·7
F	18·9	Se	49·2
Ge	90·0	Si	108·9
H	52·1	Sn	72·2
I	25·5	—	—

[a] JANAF Interim Thermochemical Tables, Dow Chemical Company, 1965.

mation is not achieved by the expression of heats of formation in terms of bond energies, because the number of compounds does not greatly exceed the number of bonds. Some bond energy terms that will prove useful in the remainder of this chapter are recorded in table 7.2. These were calculated in the sequence in which they are presented in the table, and the source compound is given with each. Enthalpies of atomization of the elements, which are given in table 7.3, and enthalpies of formation of compounds were taken from the National Bureau of Standards Technical Notes 270–1 and 270–2 unless otherwise stated.

7.4. Intrinsic bond energies.

In the last section we considered the heat evolved when four C—H bonds were formed from one gaseous carbon and four gaseous hydrogen atoms in their ground states. The choice of the ground state $1s^2 2s^2 2p^2$, ^3P, for the carbon atom is convenient but entirely arbitrary, and numerous excited states exist as alternative possibilities.

As theories of chemical bonding regard the carbon atom in methane as sp^3 hybridized, considerable theoretical interest has been focused on the excited states arising from the configuration $1s^2 2s^1 2p^3$. According to the Russell–Saunders coupling scheme, there are six, the lowest lying being ^5S with four parallel spins, and the highest, ^1P, with none. As these two terms lie 96·5 kcal/g-atom and about 340 kcal/g-atom above the ground state, our choice of one from the six for inclusion in a bond energy cycle will clearly have a big influence on the final bond energy value. Now an interesting measure of the strength of the C—H bonds in methane would be one quarter of the heat input required to remove the four hydrogen atoms and leave the four electrons on the carbon atom in the same relation to one another that they occupied in the original molecule. This quantity is often called the intrinsic bond energy, and the final state of the carbon atom 'the valence state'.

Even if we accept that the carbon atom in methane is sp^3 hybridized, the valence state does not correspond to any one of the spectroscopic terms arising from the configuration $1s^2 2s^1 2p^3$. In all six, there is spin–spin coupling between the four unpaired electrons which is destroyed when the electrons are paired in the molecule. However, by using the spectroscopically observed energies of the terms, Van Vleck allowed for this effect and calculated the degree of excitation of the sp^3 carbon valence state. This quantity is known as the promotion energy. For methane, Van Vleck obtained a figure of about 160 kcal/g-atom, so with the enthalpy of atomization of carbon in table 7.3, $\Delta H_f^0 [\mathrm{C^*(g)}]$ is about 330 kcal/mole, $\mathrm{C^*}$ referring to the valence state.[1] Thus for

$$\mathrm{CH_4(g) \to C^*(g) + 4H^{\cdot}(g)}, \quad \Delta H^0 \simeq 560 \text{ kcal/mole}, \qquad [7.5]$$

and the intrinsic bond energy, $B_I(\mathrm{C—H})$ is about 140 kcal/mole. The relation of the valence state to the Russell–Saunders terms arising from the sp^3 configuration and to the ground state is shown in fig. 7.1.

In making comparisons where promotion energies are very different a useful idea of their relative sizes may sometimes be gained from atomic spectra. In oxygen ($1s^2 2s^2 2p^4$), the terms generated by exciting a $2s$ or $2p$ electron to the $3s$, $3p$ or $3d$ orbitals lie considerably higher above the ground state than do those arising from the promotion of the valence electrons in sulphur ($1s^2 2s^2 2p^6 3s^2 3p^4$) to the $3d$ level. When both elements try to expand their octets, it is unlikely that the big difference in the two promotion energies will

---

[1] For consistency with §7.2 and 7.3, the spectroscopic promotion energy, which is an internal energy change at 0 °K (ΔE_0^0), has been equated to ΔH_{298}^0 and intrinsic bond energies calculated as enthalpies at 25°. Intrinsic bond energies, like bond dissociation energies and bond energy terms, could be expressed as ΔE_0^0 values. The small corrections required for conversion from ΔE_0^0 to ΔH_{298}^0 could be estimated by the method used in appendix 2 for terms in the Born–Haber cycle.

be offset by a sufficient difference in intrinsic bond energy, and consequently the bond energy terms in compounds such as OF_6 and OCl_4 will be much less than those in the sulphur analogues or the oxygen compounds in the $+2$ valence state. Such arguments

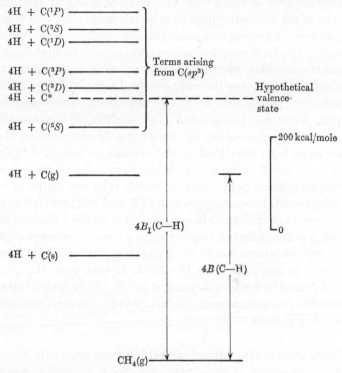

Fig. 7.1. Various types of bond energy for methane.

are believed to account for the fact that compounds such as OF_6, OCl_4, NF_5 and $F(F)_3$ do not exist.[1]

Apart from these rather qualitative applications, it is difficult to see promotion energies, and hence intrinsic bond energies, finding much quantitative use in the near future. There are several reasons for this. First, a hybridization scheme is a theoretical description of a molecule based usually on its stereochemistry. It cannot be

---

[1] More recently, an alternative approach to the same problem has used three-centre molecular orbitals, rather than d orbital hybridization schemes, to describe the bonding in these compounds. Their non-existence is then explained without considering the energies of the d orbitals.

precisely determined by experimental means and in some cases, such as the chlorine atoms in carbon tetrachloride, the stereo-chemistry gives no clue to the hybridization scheme one would adopt to describe the molecule. A further experimental barrier is the impossibility of observing the valence state in the atomic spectrum of the element when, as is usual, more than one Russell–Saunders term is generated by the valence configuration. Although the possibility of a theoretical calculation similar to that described for methane remains, even this is prevented in many of the more important cases because the experimental terms corresponding to a particular configuration have not been spectroscopically observed. Thus for phosphorus ($1s^2 2s^2 2p^6 3s^2 3p^3$) no transitions arising from the promotion of one of the $3s$ electrons to the $3d$ orbitals have been identified in the atomic spectrum. $3s^1 3p^3 3d$ is usually considered to be the hybridization scheme in molecules such as phosphorus pentachloride which take the shape of a tri-gonal bipyramid. In compounds like PCl_3 and PCl_5, where the bond energy terms are different, it is sometimes implied that an exact knowledge of the different valence state promotion energies for the two phosphorus atoms would show that the intrinsic bond energies in the two compounds were identical. If this were the case, it would of course be a telling argument for making a serious attempt to determine promotion energies, but as yet there is no evidence to suggest that it is so.

7.5. Force constants.   Bond stretching force constants, obtained from the frequencies of molecular vibrations, are sometimes used as a measure of bond strength. In polyatomic molecules, the num-ber of bond stretching and other force constants generally exceeds the number of observed frequencies, so that values cannot be calculated without introducing approximate forms for the potential energy function of the molecule. Corrections for anharmonicity are usually ignored as well [see Cottrell (1958), p. 260].

Apart from the difficulties involved in their determination, stretching force constants reflect the resistance of the bond to small perturbations. Consequently they do not always correlate with thermodynamic quantities that refer to reactions in which bonds are completely broken. The bond dissociation energy of chlorine, for example, exceeds that of fluorine, but the stretching force constants are in the reverse order. For these reasons, force constants are not used as a measure of bond strength in subsequent sections.

7.6. Some trends in bond energy terms. Of the three thermochemical measures of bond strength so far discussed in this chapter, the bond energy term is the one most easily related to the heats of formation of molecular compounds. As table 7.3 shows, the heats of atomization of many elements have been quite accurately determined, and from these and the appropriate bond energy terms, the heat of formation of a gaseous compound may be estimated if its structural formula is known. In some cases the operation is worthless because the bond energy term was originally calculated from the compound itself, but in others it offers a useful method of estimation if the assumption of additivity is a reasonable approximation. For these reasons, little emphasis is placed upon bond dissociation energies and intrinsic bond energies in the remainder of this chapter.

In §1.6, the weakness of a bond model was mentioned: while the lattice energies of ionic compounds can often be quite precisely related to internuclear distance, an analogous treatment of bond strengths does not exist. It is, however, worth commenting in a qualitative way on some interesting trends that may be discerned in bond energy terms. In Groups IV, V, VI, and VII let the element in the first period be A_1, the next A_2 and so on. Then,

(*a*) In any vertical group A_1, A_2, ..., etc., $B(A—X)$ diminishes down the series when there are no lone pairs on X. Some examples are given in table 7.4.

(*b*) When there are lone pairs on X, the bond energy order (see table 7.5) is usually

$$B(A_1—X) < B(A_2—X) > B(A_3—X) > B(A_4—X).$$

$B(A_1—X)$ is particularly small when there are lone pairs on both A_1 and X (see table 7.5).

While the second trend is not universally the case—the bond energy term in TeF_6 for instance exceeds that in SF_6, and the figure for OCl_2 is slightly less than that for OF_2—both are suggestive. Recent interest in the interpretation of bond energies has been concentrated mainly on the dissociation energies of the halogens,

TABLE 7.4 *Changes in* B(A—X) *when there are no lone pairs on* X (*kcal/mole*)

C—H	Si—H	Ge—H	Sn—H
99	76	68	60
C—C	Si—C	Ge—C	Sn—C
83	73	58	50
O—H	S—H	Se—H	Te—H
111	88	73	63

TABLE 7.5 *Changes in* B(A—X) *when there are lone pairs on* X (*kcal/mole*)

C—Cl	Si—Cl	Ge—Cl	Sn—Cl
78	96	81	75
N–F	P–F	As–F	Sb–F
67	119	116	?
N—O	P—O	As–O	Sb—O
39	88	79	?
C—N	Si—N	Ge—N	Sn—N
73	80	61	?

and in particular on the low value for fluorine. The sequence of dissociation energies $F_2 < Cl_2 > Br_2 > I_2$ has been attributed to both the repulsion between the non-bonding lone pairs on fluorine across the short internuclear distance, and to the possibility of π-bonding which involves the vacant but relatively low-lying d orbitals on chlorine, bromine and iodine. Some stereochemical evidence for this type of bonding has been found in molecules such as trisilylamine (see below).

These two phenomena are consistent with trend (ii), the difference, for example, between the C—Cl and Si—Cl bond energies being due to the high energy of the $3p$ and $3d$ orbitals on carbon which inhibits any $p_\pi \rightarrow p_\pi$ or $p_\pi \rightarrow d_\pi$ bonding. The small O—O, N—O and N—N bond energies are due to a combination of this effect with the repulsion between non-bonding lone pairs. When these two effects are imposed upon a normal decrease with inter-nuclear distance, they may possibly be the principal cause of ir-regularities in the variations in bond energy terms down a group. Nevertheless, the evidence for this argument is only circumstantial and it should be regarded as speculative.

The possibility of p_π—d_π bonding emphasizes a particularly important point. In table 7.2, the bond order corresponds to a number of lines drawn between two atoms which is equal to the number of pairs of electrons assigned to the bond. This pheno-menon would mean that the Si—Cl 'single-bond', for example, is not a single bond at all. The standard bonds of integral multiplicity are selected only with the aid of a particular electronic theory of valency. Bond order cannot be measured experimentally and only has a meaning in relation to this theory.

7.7. Electronegativities. There exists a rough method for the calculation of bond energy terms for 'single' bonds which is due to the work of Pauling. He showed that a constant x could be assigned to each of the non-metals such that approximately:

$$(x_A - x_B)^2 = \tfrac{1}{23}(B(A\text{—}B) - \sqrt{[B(A\text{—}A) \cdot B(B\text{—}B)]}), \qquad (7.10)$$

where x_A and x_B are the constants for two different non-metals, and bond energy terms are in kcal/mole. (7.10) is a later development of an earlier formula which contained the arithmetic instead of the geometric mean of $B(A\text{—}A)$ and $B(B\text{—}B)$ in the bracket on the right. Substitution of the geometric mean was found to give positive values for the right-hand side where in a number of previous cases the figure had been negative.

The constants x were called by Pauling 'electro-negativities' because he considered them a measure of 'the power of an atom to attract electrons to itself.' Partly because of this, the possibility of using them to calculate unknown bond energy terms has received relatively little attention. The values recorded in table 7.6 were assigned by an appraisal of the bond energy terms in table 7.2; (7.10) was used throughout. In column three of table 7.2, the cal-culated values of the bond energy terms are compared with the experimental figures. Although in some cases, notably I—F, F—O, N—H and N—F, the agreement is poor, the average dis-

TABLE 7.6   *Values of x calculated from* (7.10)

Element	x	Element	x
As	2·20	N	3·15
B	1·90	O	3·65
Br	3·05	P	2·20
C	2·45	S	2·75
Cl	3·25	Sb	2·10
F	4·00	Se	2·55
Ge	2·20	Si	1·95
H	2·20[a]	Sn	2·10
I	2·80	—	—

[a] Values refer to a scale on which $x_H = 2·20$.

crepancy for some sixty bonds is less than 3 kcal/mole. In all cases where a comparison of calculated and experimental values was made, only one valency of an element, A, was considered, and this was the same in both the bond A—B and the bond A—A. It is sometimes stated that x_A varies with the valency of A because $B(A—B)$ does so, but in quoted examples an attempt is not usually made to ensure that the value of $B(A—A)$ is that appropriate to the new valency. Whether this last operation would keep x constant is not known, for sufficient thermochemical data is not yet available. One thing is certain: in using (7.10) and the figures in table 7.6 to calculate bond energy terms, the valency of A or B should not be different in $B(A—B)$ and $B(A—A)$ or $B(B—B)$. This may be profitably illustrated by an example. The bond energy terms for the molecules SF_4 and SF_6 are 82 and 79 kcal/mole respectively, giving, with the value of $B(S—S)$ for divalent sulphur, figures of 2·80 and 2·86 for x respectively. Using these values for S^{IV} and S^{VI}, the calculated bond energy terms for SH_4 and SH_6 are 89 and 91 kcal/mole. These figures yield -81 kcal/mole and -167 kcal/mole for $\Delta H_f^0[SH_4(g)]$ and $\Delta H_f^0[SH_6(g)]$. As $\Delta H_f^0[SH_2(g)]$ is $-4·9$ kcal/mole, the calculation suggests that the two unknown compounds are exceedingly stable both with respect to their constituent elements and to H_2S. While the possibility of their existence should not be discounted, it does seem unlikely that they are anywhere near as stable as the calculation implies. As H_2S can be prepared by passing hydrogen through boiling sulphur, it would be surprising if kinetic factors prevented the uptake of further hydrogen when the latter was in excess. If (7.10) is used to calculate bond energy terms in subsequent sections, the operation will be carried out with the elements in the valencies in which they occur in table 7.2. Attempts will then be made to see whether the results agree with known chemical facts.

7.8. Introduction to the remaining sections. In the remaining sections of this chapter, a few aspects of the chemistry of some non-metals are discussed in a way that tries to bring out the relation between chemical and structural properties and bond strengths. The heats of formation of some hypothetical compounds are also calculated and the possible reasons for their instability discussed. Owing to shortage of space, the discussion of individual reactions and of structures of compounds is extremely terse, and textbooks such as Cotton and Wilkinson's *Advanced Inorganic Chemistry* (1966), and Well's *Structural Inorganic Chemistry* (1962) will be found very useful in following the argument.

TABLE 7.7 *Inverse correlations between bond length and bond strength*

Compound	C—O or N—O bond length (Å)	Corresponding bond energy term (kcal/mole)
Alcohols	1·42	86
Aldehydes	1·22	176
CO_2	1·16	192
CO	1·13	257
NH_2OH	1·46	39
NO_2	1·20	112
NO	1·15	151

Additivity of bond energy terms within the particular valencies of an element will be assumed throughout. It must be emphasized that this is only an approximation, but because departures from it are often small compared with the heats or variations in the heats of many interesting chemical reactions, the assumption is a very useful one. Attempts to relate the equilibrium constants of reactions of non-metallic compounds to bond energy terms ignore any free energies of condensation of reactants or products, and usually need to assume that ΔS^0 or the variations in ΔS^0 for the gas phase reaction are small compared with ΔH^0 or the variations in ΔH^0. These approximations are mentioned frequently in the text but should be marked by the reader.

When the bond energy term for the link between the same two atoms varies *considerably*, there is usually an inverse correlation with bond length. Some examples for carbon–oxygen and nitrogen–oxygen bonds are shown in table 7.7. In structural theory, of course, this variation is attributed to changes in bond order. In discussing the structures of compounds, the inverse correlation will frequently be demonstrated as an empirical principle.

Unless 'bond dissociation energy' or 'intrinsic bond energy' are referred to by name, all comments on the strengths or energies of chemical bonds refer to bond energy terms.

7.9. Boron. The bond dissociation energy of the gaseous molecule BO, which can be produced by heating boron and B_2O_3 to $1000°$, is 188 kcal/mole. The simple molecular orbital theory for diatomic molecules suggests that the bond in this compound should have an order of 2·5, so the strength of any multiple link between boron and oxygen is unlikely to be more than twice that of the very strong single bond (125 kcal/mole). This is reflected in the stereochemistry of B_2O_3 and the borates which are the naturally occurring forms of the element, and consist of extended arrays of B—O—B single bonds formed usually from triangular BO_3 or tetrahedral BO_4 groups [see fig. 7.2 and Wells (1962)]. Such structures are thermodynamically preferable to those involving chain termination by the formation of boron–oxygen multiple bonds $(O{=}B{-}O^-,\ O{=}B{-}O{-}B{=}O)$. Glassy B_2O_3 consists of a randomly oriented, three dimensional network of BO_3 groups in which each boron is bound to three oxygens and each oxygen to two borons. Although the enthalpy of sublimation of solid boron is rather large, the formation of three very strong B—O single bonds easily offsets both this and the heat of atomization of three oxygen atoms, and the free energy of formation of B_2O_3 is -285 kcal/mole. The production of boron from this and naturally occurring borates is therefore a difficult process, and the action of powerful reducing agents like potassium and magnesium is necessary to extract the element.

Although the four trihalides of boron are monomeric, appreciable quantities of the corresponding hydride, BH_3, have not been prepared. The value of the B—H bond energy term obtained from (7.10) and the figures in tables 7.2 and 7.6 is 89 kcal/mole which

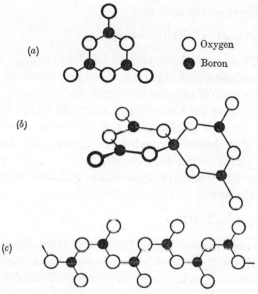

Fig. 7.2. Some borate anions: (a) The $B_3O_6^{3-}$ ion in $K_3B_3O_6$. (b) The $B_5O_{10}^{5-}$ ion in $KH_4B_5O_{10}.2H_2O_4$. (c) Portion of the infinite $B_2O_4^{2-}$ chain in CaB_2O_4.

yields $\Delta H_f^0[BH_3(g)] = +24$ kcal/mole. With this, and the known figure for $\Delta H_f^0[B_2H_6(g)]$, ΔH^0 for the reaction,

$$2BH_3(g) \rightarrow B_2H_6(g) \tag{7.6}$$

is about -40 kcal. This figure is in good agreement with a recent experimental value. Assuming that the entropy of $BH_3(g)$ is the same as that of $NH_3(g)$, $\Delta S^0 = -37$ cal/deg. and $\Delta G^0 = -30$ kcal at $25°$. Borine is evidently unstable at room temperature to B_2H_6, and the loss of two B—H bonds in the course of reaction must be more than compensated by the formation of the two unusual, but now well-established, B—H—B three-centre bonds. This compensation is about 20 kcal/mole per bond formed, so from the value of B(B—H) above, the 'bond energy term' for B—H—B must be around 110 kcal/mole. This means that the mean bond energy per hydrogen atom in B_2H_6 is close to B(C—H). In addition, the B—O bond energy term is very nearly two thirds of B(C=O) in CO_2. Taken alone, these facts suggest that the reactions,

$$B_2H_6(g) + 3O_2(g) \rightarrow B_2O_3(s) + 3H_2O(g) \qquad [7.7]$$

for which $\Delta H^0 = -486$ kcal and

$$C_2H_6(g) + \tfrac{7}{2}O_2(g) \rightarrow 2CO_2(g) + 3H_2O(g) \qquad [7.8]$$

for which $\Delta H^0 = -341$ kcal, should be about equally exothermic, but in the combustion of ethane a C—C link must be broken and an extra half a mole of oxygen atomized. Consequently one mole of diborane burns with the evolution of some $(83 + 60)$ kcal more heat than ethane. In view of this fact, the current interest in the possibility of using boron hydrides as lightweight fuels of high calorimetric yields is understandable.

The boron halides form molecular addition compounds by reactions of the type,

$$NR_3 + BX_3 \rightarrow R_3N.BX_3 \qquad [7.9]$$

in which the stereochemistry of the boron atom changes from triangular to tetrahedral. Considerable research has established that, with a particular Lewis base, the order of acceptor power is $BBr_3 > BCl_3 > BF_3$. Some thermochemical evidence for this observation is shown in table 7.8 where the heats of reaction of the three halides with a solution of pyridine in nitrobenzene are recorded:

$$C_5H_5N(soln) + BX_3(g) \rightarrow C_5H_5N.BX_3(soln). \qquad [7.10]$$

If the difference in the heats of solution of the three addition compounds are small, then their bond dissociation energies:

$$C_5H_5N.BX_3(g) \rightarrow C_5H_5N(g) + BX_3(g) \qquad [7.11]$$

should vary in the same order as the heats of reaction in table 7.8.

In other cases it is usual to argue that the bond formed is strengthened by the attraction of the groups attached to the boron atom for the lone pair of electrons on the base and weakened by the repulsive forces between these same groups when they are pushed

TABLE 7.8   *Enthalpies of the reaction:*
$C_5H_5N(\text{soln}) + BX_3 (\text{g}) \rightarrow C_5H_5N . BX_3 (\text{soln})$ *in nitrobenzene (kcal)*

BF$_3$	BCl$_3$	BBr$_3$
$-34\cdot2$	$-45\cdot2$	$-51\cdot8$

TABLE 7.9   *Enthalpies of reaction of borine derivatives with tri-methylamine in the gas phase (kcal)*

BH$_3$	BH$_2$(CH$_3$)	BH(CH$_3$)$_2$	B(CH$_3$)$_3$
-36	?	$-23\cdot6$	$-17\cdot6$

closer together in the addition compounds. Such an argument can account for the order of acceptor power of the methyl borines shown in table 7.9. As the introduction of bulky methyl groups should increase the repulsion term and as, relative to hydrogen, the methyl group is thought to be an electron donor, this is

$$BH_3 > BH_2CH_3 > BH(CH_3)_2 > B(CH_3)_3.$$

With the halides however, the observed sequence is the reverse of that expected, for fluorine is the most electronegative and least bulky of the halogens. To account for this result, the lone pairs of the halogens in the planar BX$_3$ molecules are assumed to donate to the vacant unhybridized p orbital on the boron. This effect is the source of an additional contribution to the binding which is lost when the boron atom is transformed into tetrahedral coordination. If it is assumed that the extent of this π-bonding is much diminished as the halogen becomes larger, then the ease of formation of the tetrahedral configuration will run in the order BF$_3$ < BCl$_3$ < BBr$_3$ and generate the same order of acceptor power.

Unexpectedly short bonds in BF$_3$ add some support to this concept. In (CH$_3$)$_3$N.BF$_3$, H$_3$N.BF$_3$ and (CH$_3$)$_2$O.BF$_3$ where no such π-bonding is possible, the B—F bond lengths are 1·39, 1·38 and 1·43 Å, while in BF$_3$ itself they are 1·30 Å.

Similar increases in the boron–oxygen bond lengths occur when boron changes from planar to tetrahedral coordination. Structural determinations of a number of borates containing both types show that the bond lengths in the BO_3 triangles are about $1 \cdot 37$ Å while those in the BO_4 tetrahedra are close to $1 \cdot 48$ Å. That both types of coordination occur widely suggests that the triangular bonds are stronger than the tetrahedral, for a larger number of bonds would otherwise ensure the predominance of the latter. These facts suggest that the outstanding thermochemical feature of boron chemistry, the strength of the boron–oxygen 'single' bond in BO_3 triangles, may be partly due to $p_\pi \to p_\pi$ bonding from the lone pairs on oxygen to the vacant p orbital on boron.

7.10. Carbon and silicon. The classical formulation of the structure of graphite would have shown each carbon atom in the layers forming one C=C and two C—C bonds. As $B(C{=}C)$ is considerably less than twice $B(C{-}C)$, this, by itself, suggests that the allotrope would be unstable with respect to diamond, but the resonance energy attributed to electron delocalization within the layers and the van der Waals forces between them both combine to make diamond the metastable form by a fraction of a kilocalorie.

Silicon only occurs as a diamond form and no compounds containing the Si=Si bond are known. No reliable value for its energy can be estimated, but it is interesting to note that the bond dissociation energy of C_2 which, according to simple M.O. theory, has a bond order of two, is almost exactly $B(C{=}C)$. The assumption that this holds for silicon too gives $B(Si{=}Si) = 76$ kcal/mole and, with the value of $B(Si{-}H)$ in table 7.2, $\Delta H_f^0 [Si_2H_4] = +46$ kcal/mole. Decomposition reactions into the elements or silicon and monosilane or polymers of the type $(SiH_2)_n$ would probably be faster than the corresponding ethylene reaction because any bonds broken in a rate-determining step would be weaker.

Both carbon and silicon form saturated hydrides with the formula A_nH_{2n+2}. In diamond and silicon, each atom is tetrahedrally coordinated to four others, so the rupture of four bonds liberates two atoms of the elements. Thus in all reactions of the type,

$$A_{n-1}H_{2n}(g) + A(s) + H_2(g) \to A_nH_{2n+2}(g), \qquad [7.12]$$

where A is in the diamond form, one H—H and two A—A bonds are broken, while one A—A and two A—H bonds are formed.

Consequently, to a first approximation which assumed additivity of bond energies and ignores the heats of condensation of the hydrides, the standard enthalpy of formation per mole of hydrogen is the same for every silane or every paraffin. As monosilane is endothermic and, owing to the moles of hydrogen gas consumed in the formation of silanes, the standard entropy of formation is negative, all the silanes are thermodynamically unstable with respect to their constituent elements at room temperature. They do in fact decompose in this way when they are heated to 400–500°.

It is interesting that the weakness of the Si—Si bond tends to stabilize the silanes to this reaction because there are more Si—Si bonds in the silicon produced than in the parent silane. The decomposition into the elements is favoured by the comparatively high enthalpy of atomization of hydrogen and the comparatively low Si—H bond energy. When chlorine, an element that has a lower enthalpy of atomization and forms stronger bonds with silicon, is substituted for hydrogen, the stability with respect to the elements is much increased. Compounds of the formula, Si_nCl_{2n+2}, are known with n as high as ten, and in cases so far studied, the formation of silicon and chlorine is not observed below 800°. On the other hand, disproportionation reactions such as,

$$5Si_2Cl_6 \rightarrow 4SiCl_4 + Si_6Cl_{14} \tag{7.13}$$

and

$$nSi_2Cl_6 \rightarrow nSiCl_4 + (SiCl_2)_n \text{ (solid polymer)} \tag{7.14}$$

in which the number and type of each bond remains constant, occur rather readily.

Kinetic factors appear to be of paramount importance in the pyrolysis of the paraffins. At 600°, decomposition to the elements is thermodynamically favourable, but the industrially important 'cracking' processes are the dominant reaction.

Thermodynamics draws a sharp distinction between the stability of silanes and paraffins with respect to hydrolysis:

$$M_nH_{2n+2} + 2nH_2O \rightarrow nMO_2 + (3n+1)H_2. \tag{7.15}$$

In silica the total energy of the four bonds formed by each silicon exceeds that of the two multiple bonds in carbon dioxide, while $B(M—M)$ and $B(M—H)$ are both lower for the second row element.

All these facts favour hydrolysis of the silanes relative to the paraffins, and in the presence of a trace of base, the former are decomposed by water. By contrast, the paraffins are thermodynamically stable to hydrolysis. On the other hand most organic compounds, including the paraffins, burn if heated or ignited but the silanes are spontaneously inflammable in oxygen at normal temperatures. The difference in the stabilities of silicon and carbon chains to combustion must therefore lie in the kinetics of the reaction.

The combination of kinetic and thermodynamic effects in these last two reactions mainly accounts for the apparent difference in the catenating abilities of carbon and silicon. The instability of the Si—Si bond with respect to combustion and hydrolysis ensures that the earth's silicon content exists as oxidized forms of the element rather than catenated compounds.

Nucleophilic substitution is another field where kinetics often differentiates sharply between carbon and silicon. In cases where it is thought likely that an S_N2 mechanism is involved, the availability of the $3d$ orbitals for the formation of five-coordinate intermediates is often advanced as a reason for the fast decomposition of silicon compounds. The hydrolysis of both the carbon and silicon tetrahalides is thermodynamically feasible, but only the silicon compounds are decomposed by water:

$$CCl_4\,(l) + 2H_2O \rightarrow CO_2 + 4HCl\,(aq), \quad \Delta G^0 = -91\,\text{kcal}, \qquad [7.16]$$

$$SiCl_4\,(l) + 2H_2O \rightarrow SiO_2 + 4HCl\,(aq), \quad \Delta G^0 = -69\,\text{kcal}. \qquad [7.17]$$

It is particularly interesting to compare the rapid hydrolysis of structure I with the inertness of the carbon analogue. When M = C,

I

the cage-like structure inhibits rearward attack by hydroxide ions while the expulsion of chloride is hindered by the resistance to the production of a planar carbonium ion. Consequently neither S_N1 or S_N2 reactions are favoured. When M = Si however, it is claimed that the availability of the $3d$ orbitals allows the formation of a five-coordinate intermediate by frontal attack. Certainly the hydrolysis of I occurs readily without inversion of configuration.

It is probably in the compounds that they form with oxygen that carbon and silicon display their most striking differences. The mean bond energy in carbon dioxide is 192 kcal/mole, some 15 kcal/mole greater than the value for C=O in aldehydes and ketones, while the internuclear separation of the carbon and oxygen atoms is 1·16 Å compared with 1·22 Å in acetaldehyde and acetone, and 1·13 Å in the triply bound CO molecule. Pauling attributed the strengthening of the bond to resonance contributions from the structures $\overset{+}{O}{\equiv}C{-}\overset{-}{O}$ and $\overset{-}{O}{-}C{\equiv}\overset{+}{O}$.

Because the bond energy term in CO_2 is more than twice $B(C{-}O)$, a silica-like structure is thermodynamically unstable to the discrete gaseous molecule. In an extended array of C—O bonds, the rupture of four linkages requires 342 kcals and liberates one carbon and two oxygen atoms. For the process,

$$C(s) + O_2(g) \rightarrow C(g) + 2O(g), \quad \Delta H^0 = +290 \text{ kcal} \qquad [7.18]$$

so $\Delta H_f^0 [CO_2 (\text{silica form})] = -52$ kcal/mole. As

$$\Delta H_f^0 [CO_2(g)] = -94 \text{ kcal/mole},$$

and the entropy of sublimation of the silica form would be positive, the isolation of the condensed place in the favourable reaction:

$$CO_2(\text{silica form}) \rightarrow CO_2(g), \quad \Delta H^0 = -42 \text{ kcal/mole}, \qquad [7.19]$$

is only a remote possibility. For similar reasons, oxyanions of carbon with extended structures are unstable with respect to the simple carbonate ion.

The discrete gaseous nature of the oxides of carbon is a property of huge industrial importance. The formation of metallic oxides from the elements at ordinary temperatures occurs with the uptake of a gas, so the reaction is attended by a large entropy decrease. Furthermore, the boiling point of the metal is often considerably lower than that of the oxide, so once boiling has occurred, this decrease becomes even larger. From (3.1) therefore, the free energies of formation of metallic oxides must rapidly become more positive when the temperature is raised. Carbon, however, possesses a unique combination of properties: the element boils at very high temperatures and forms two exothermic gaseous oxides, CO and

CO_2. At normal temperatures, CO_2 is the more stable, but of the two reactions,

$$C(s) + O_2(g) \rightarrow CO_2(g), \quad \Delta G^0 = -94 \cdot 3 \text{ kcal/mole}, \qquad [7.20]$$

$$C(s) + \tfrac{1}{2}O_2(g) \rightarrow CO(g), \quad \Delta G^0 = -32 \cdot 8 \text{ kcal/mole}, \qquad [7.21]$$

only the second occurs with a change in the number of moles of gas and a large entropy increase. As the temperature is raised, $\Delta G_f^0[CO(g)]$ rapidly becomes more negative, and above 710°, where the standard free energy change of the reaction,

$$C(s) + CO_2(g) \rightarrow 2CO(g) \qquad [7.22]$$

is zero, the monoxide is the more stable. Above 710°, $\Delta G_f^0[CO(g)]$ continues to decrease while that of metallic oxides increases. Consequently carbon, which is both cheap and plentiful, will reduce almost any metallic oxide if the temperature is sufficiently raised. Thus at 25°, $\Delta G^0 = 103 \cdot 3$ kcal for the reaction,

$$C(s) + MgO(s) \rightarrow Mg(s) + CO(g) \qquad [7.23]$$

but the metal boils at 1120° and above 1900° the oxide is reduced to magnesium vapour. The major technical difficulty of this process, which was used for the wartime production of the metal, is the need for very rapid cooling of the products to prevent reversal of the reaction. In extraction processes, the gaseous nature of carbon monoxide is also important for kinetic reasons. The reaction between solid carbon and solid oxides is necessarily slow, but gas–solid reactions of the type,

$$2CO + SnO_2 \rightarrow 2CO_2 + Sn \qquad [7.24]$$

are faster. Carbon monoxide can therefore act as an intermediate, the carbon dioxide produced being reduced via [7.22] and the overall process becoming

$$2C + SnO_2 \rightarrow Sn + 2CO. \qquad [7.25]$$

Thus although the reduction process is favourable at lower temperatures, it is usually preferable to operate above 700° where large concentrations of carbon monoxide are present.

Although many metals occur as sulphide ones, CS_2 is an endothermic compound and carbon monosulphide is unstable at furnace temperatures. Consequently the direct reduction of sulphides by carbon has virtually no industrial applications, and the sulphide ores of metals such as zinc, lead, antimony and arsenic are normally converted to the oxides by roasting in air and then reduced with anthracite.

The oxygen chemistry of silicon is very complex, but the many types of structures found in silica and the silicates are all made up of SiO_4 tetrahedra linked in a variety of different ways. There is little doubt that silicon can form strong multiple bonds with oxygen, but compounds containing them are unstable with respect to others containing the very strong single bond. Thus, at 153 kcal/mole, $B(Si=O)$ in gaseous SiO_2 exceeds the dissociation energy of nitric oxide, but this is nowhere near double $B(Si—O)$ and so the latent heat of sublimation of silica is about 140 kcal/ mole. Again, the bond dissociation energy of SiO is 192 kcal/ mole. This is less than that of carbon monoxide but the lower value of the enthalpy of atomization of the element results in very similar enthalpies of formation. However, for the reaction

$$2SiO\,(g) \to Si\,(s) + SiO_2\,(s) \tag{7.26}$$

$\Delta H^0 = -170$ kcal, and SiO can only be obtained by taking advantage of the van't Hoff isochore, and heating silicon and silica to about 1500°.

As with boron, the outstanding thermodynamic feature of silicon chemistry is the strength of the 'single' bond that it forms with oxygen. Again this may be due to the fact that it is not a single bond at all. The relative availability of the d orbitals on a second row element which was mentioned earlier, means that the oxygen lone pairs might be used in $(p_\pi \to d_\pi)$ bonding of the type discussed in §7.6. The difficulty of defining a standard single bond, which at present is a theoretical question, is a formidable barrier to the detection of this type of bonding. The major experimental evidence is found in the unusual stereochemistry of the trisilylamine molecule where the nitrogen and the three silicons lie in the same plane. In the planar configuration, it is thought that strong overlap of the unhybridized p lone pair with the vacant d orbitals of the three silicons is achieved (see fig. 7.3) and the binding is strengthened. This should provide an additional barrier to any reaction in which the lone pair is donated to a Lewis acid, and the coordination changes to tetrahedral. In agreement with this idea, trisilylamine is a much weaker base than the pyramidal molecule of trimethyl-amine.

In a similar way, the Si—O—Si angle of 144° in $O(SiH_3)_2$ could be cited as evidence of silicon–oxygen ($p_\pi \to d_\pi$) bonding, for the corresponding figure of 111° in dimethyl ether is close to the expected tetrahedral angle. In any case, since oxygen possesses two lone pairs, one of which could be said to be perpendicular to the Si—O—Si plane, it is apparent that, on symmetry grounds,

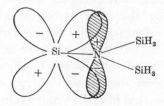

Fig. 73. Orbital overlap for the formation of a possible
($p_\pi \to d_\pi$) bond in trisilylamine.

considerable $p_\pi \to d_\pi$ bonding is possible whatever the angle. Both the bond angle at oxygen in disilyl ether and the Si—O distance of 1·63 Å are very similar to the values in the crystoballite and quartz forms of silica from which the value of $B(Si—O)$ is usually calculated.

The Si—N bond energy in table 7.2 was calculated from $\Delta H_f^0 [Si_3N_4(s)]$. In this compound, the nitrogen displays planar three-fold coordination with an Si—N distance almost identical with that in $N(SiH_3)_3$. $B(Si—N)$, which is one-twelfth the enthalpy of atomization of $Si_3N_4(s)$, exceeds $B(C—N)$ by quite a small margin, while the difference between $B(Si—O)$ and $B(C—O)$ is over 20 kcal/mole. Thus the bond energy terms give no indication that the bonds formed by planar nitrogen with silicon have a contribution that is absent in those formed by oxygen.

There is a growing body of *circumstantial* evidence that ($d_\pi \to p_\pi$) bonding is quite common, and that the lack of 'multiply' bonded compounds containing second row elements is due as much to the strength of the 'single' bonds in which it operates as to an unusual weakness of the 'multiple' bonds. The extended nature of silicate structures which are based on SiO_4 tetrahedra, is a particularly good example of this.

7.11. Nitrogen and phosphorus. While $B(N{\equiv}N)$ greatly exceeds $B(P{\equiv}P)$ in the diatomic molecules, $B(N{-}N)$ is less than $B(P{-}P)$. These facts are reflected in the nature of the naturally occurring forms of the elements, for white phosphorus consists of tetrahedral P_4 molecules in which the atoms are held together by six P—P single bonds. It is apparent from the low standard enthalpy of the process:

$$4P\,(\text{white}) \to P_4(g), \quad \Delta H^0 = +14 \cdot 1 \text{ kcal} \qquad [7.27]$$

that the tetrahedra themselves are bound together in the crystal only by van der Waals forces. From the value of $B(N{-}N)$ in table 7.2, the estimated value of $\Delta H_f^0[N_4(g)]$ is $+224$ kcal/mole, so the molecule, or a solid containing it, would be very unstable to diatomic nitrogen. By contrast, for the reaction:

$$P_4(g) \to 2P_2(g), \quad \Delta H^0 = +54 \cdot 9 \text{ kcal.} \qquad [7.28]$$

The great difference in the strengths of the single and triple nitrogen–nitrogen bonds accounts for the unwillingness of the element to undergo catenation. From the bond energies in table 7.2 we obtain for the compound $H_2N(NH)_{n-2}NH_2(g)$:

$$\Delta H_f^0 = (34n{-}44) \text{ kcal/mole.} \qquad (7.11)$$

This suggests that these compounds are all endothermic for $n \geqslant 2$ and, as the entropies of decomposition are favourable, that they are thermodynamically unstable to nitrogen and hydrogen and to nitrogen and ammonia. Hydrazine decomposes in a combination of these ways when it is heated to 250°, while the addition of platinum black to a solution of the compound brings about catalytic decomposition to aqueous ammonia and nitrogen at room temperature. It has, of course, been assumed in these arguments that the heats of solution or condensation of hydrazine and other chain compounds are relatively small.

Hydrazine also forms molecular nitrogen when it burns in air, and in the presence of platinum will explode in pure oxygen at only 30°. Alternatively, hydrogen peroxide can be used as a source of oxygen. This takes advantage of the weakness of the O—O as well as the N—N single bond and adds twice the heat of decomposition of hydrogen peroxide to the normal energy yield of the simple combustion:

$$N_2H_4(l) + 2H_2O_2(l) \rightarrow N_2(g) + 4H_2O(g), \quad \Delta H^0 = -153\cdot5 \text{ kcal.}$$

$$[7.29]$$

The reaction affords a means of obtaining high yields of hot gases from reactants of low volume. Concentrated hydrazine and hydrogen peroxide solutions formed the basis of an early rocket fuel.

If the low dissociation energy of fluorine is really due to repulsions between non-bonding electrons, then the same phenomenon probably accounts for the low values of $B(N\text{—}N)$ and $B(N\text{—}O)$. Compounds containing the N—O single bond tend to be unstable with respect to nitrogen oxides made up of the strong multiple bonds that the elements can form with one another, or to nitrogen and oxygen. Thus above 15°, hydroxylamine decomposes by two separate routes:

$$3NH_2OH \rightarrow NH_3 + N_2 + 3H_2O, \qquad [7.30]$$

$$4NH_2OH \rightarrow 2NH_3 + N_2O + 3H_2O, \qquad [7.31]$$

while above 0°, N_2O_5 dissociates thus:

$$O_2N\text{—}O\text{—}NO_2 \rightarrow O_2N\text{—}NO_2 + \tfrac{1}{2}O_2. \qquad [7.32]$$

From the bond energies in table 7.2, the heat of formation of N_4O_6, the nitrogen analogue of phosphorus trioxide, may be calculated. This molecule consists of a tetrahedron of nitrogen atoms linked only by bridging oxygens, and contains twelve N—O single bonds.

$$\Delta H_f^0[N_4O_6(g)] \simeq 4 \times 113\cdot0 + 6 \times 59\cdot6 - 12 \times 39, \qquad (7.12)$$

$$> +300 \text{ kcal/mole}, \qquad (7.13)$$

and for the reaction,

$$N_4O_6(g) \rightarrow 2N_2O_3(g), \qquad [7.33]$$

$$\Delta H^0 < -260 \text{ kcal.}$$

Thus the future isolation of this unknown compound is very unlikely.

Again for the unknown acid, $N(OH)_3$, we estimate

$$\Delta H_f^0 [N(OH)_3(g)] = -2 \text{ kcal/mole}$$

and for the reactions

$$2N(OH)_3(g) \rightarrow N_2(g) + \tfrac{3}{2}O_2(g) + 3H_2O(l) \qquad [7.34]$$

and

$$N(OH)_3(g) \rightarrow HNO_2(aq) + H_2O(l), \qquad [7.35]$$

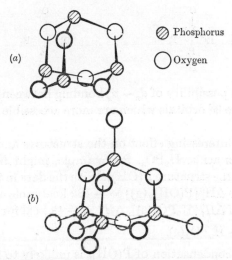

(a)

⊘ Phosphorus

◯ Oxygen

(b)

Fig. 7.4. Structures of (a) P_4O_6 and (b) P_4O_{10} in the vapour phase.

$\Delta H^0 = -203$ kcal and -95 kcal respectively. Clearly the unknown but obviously small heat change associated with the condensation of the compound into liquid, solution or crystalline form would not offset the very favourable enthalpies of decomposition.

Both nitrogen and phosphorus form compounds of the type X_3PO and X_3NO, nitrogen in the amine oxides and NOF_3, and phosphorus in the phosphine oxides and phosphoryl halides, but there is evidence that the bonds to oxygen are different in the two cases. The N—O bond length in $(CH_3)_3NO$ is 1·44 Å and hardly differs from the N—O distance in hydroxylamine, but the phosphorus-oxygen separation in $POBr_3$, for example, is 1·41 Å compared with 1·62 Å in P_4O_6. A direct comparison of the two types is possible in P_4O_{10}. This differs from P_4O_6 (see fig. 7.4, from Wells, 1962) only in the presence of four additional apical oxygen atoms each linked to only one phosphorus. The apical internuclear separation is 1·39 Å compared with 1·62 Å in the bridge. These facts suggest that the apical bonds in molecules like $POCl_3$, P_4O_{10}, and the phosphorus acids possess some multiple bond character. Consequently it is customary to write the amine oxide formula with a classical dative bond and the phosphorus compounds with a P=O double bond (see II and III). The difference is usually

$$O^-$$
$$|$$
$$R\diagdown \overset{N^+}{\underset{|}{}} \diagup R$$
$$R$$

II

$$O$$
$$||$$
$$R\diagdown \overset{P}{\underset{|}{}} \diagup R$$
$$R$$

III

ascribed to the possibility of d_π—p_π bonding between the oxygen p orbitals and the $3d$ orbitals which are more accessible in the second row element.

This has an interesting effect on the structures and basicities of the phosphorus acids. H_3PO_3, for example, might have been expected to have the structure $P(OH)_3$. With the data in tables 7.2 and 7.3 we estimate $\Delta H_f^0[P(OH)_3(g)] = -182$ kcal/mole and, using the known value of $\Delta H_f^0[H_3PO_3(s)]$, $\Delta H^0 = -49$ kcal for the reaction,

$$P(OH)_3(g) \rightarrow H_3PO_3(c).\qquad\qquad\qquad [7.36]$$

As the heat of condensation of $P(OH)_3$ is unlikely to be as large as 50 kcal/mole, the estimate is in accord with the non-existence of this compound. By removing one of the hydrogen atoms and linking it to phosphorus, the number of bonds formed by the element is increased, and a more stable dibasic acid, $H(OH)_2P=O$, is formed. Although the energy of the O—H bond that is broken almost certainly exceeds that of the P—H bond that is formed, the additional strength of the apical bond more than compensates for this difference. Similarly hypophosphorous acid is monobasic $H_2(OH)P=O$ rather than dibasic $HP(OH)_2$.

The varied structures of the condensed phosphates are built up of the basic unit IV which occurs in phosphoric acid and in the

$$O$$
$$||$$
$$O\diagup \overset{P}{\underset{|}{}} \diagdown O$$
$$|$$
$$O$$

IV

orthorhombic form of P_2O_5. Negative charges may be written on the oxygen atoms when one or more of the three chains are blocked, and the equality of the bond lengths between phosphorus and apical oxygens can be interpreted by invoking a resonance effect. Some examples of such structures are shown in fig. 7.5.

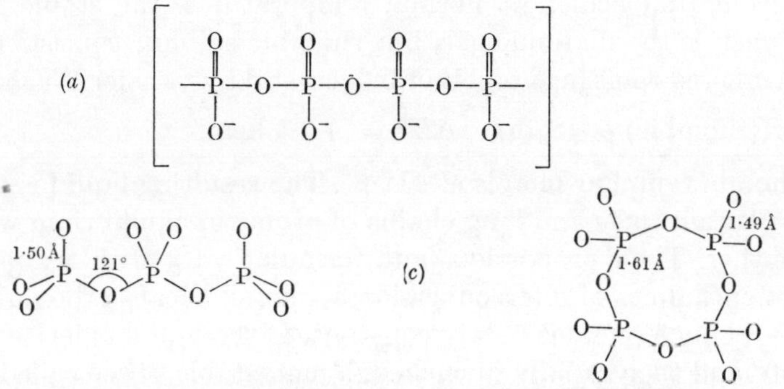

Fig. 7.5. Condensed phosphates: (a) Linking of PO_4 tetrahedra to give the long chain anion in sodium metaphosphate glass. (b) The $P_3O_{10}^{5-}$ ion in $Na_5P_3O_{10}$. (c) The $P_4O_{12}^{4-}$ ion in $(NH_4)_4P_4O_{12}$.

The phosphorus halides are all instantly hydrolysed by water. PCl_3 decomposes according to the equation,

$$PCl_3 + 3H_2O \rightarrow H_3PO_3 + 3HCl, \qquad [7.37]$$

while nitrogen trichloride is slowly hydrolysed to form ammonia and hypochlorous acid:

$$NCl_3 + 3H_2O \rightarrow NH_3 + 3HClO. \qquad [7.38]$$

This difference is usually attributed to the absence of relatively low lying d orbitals on nitrogen for the coordination of hydroxide ions in a reaction intermediate. The nucleophile is forced to attack the chloride ions instead, although the decomposition to nitrous acid and HCl is thermodynamically more favourable. It is worth noting that this mode of hydrolysis is not possible for CCl_4, for $B(C-Cl)$ is much greater than $B(N-Cl)$, while the difference in the energies of the bonds formed with hydrogen is small:

$$CCl_4 (l) + 4H_2O (l) \rightarrow CH_4 (g) + 4HClO (aq), \quad \Delta G^0 = +154 \text{ kcal.}$$

$$[7.39]$$

7.12. Oxygen and sulphur.   The dissociation energies of O_2 and S_2 are 119 and 103 kcal/mole, but the corresponding single bond energies are 34 and 63 kcal/mole. Thus, in contrast to the behaviour of oxygen, chains or rings of sulphur atoms connected by single bonds are thermodynamically stable to the formation of the diatomic molecule. At normal temperatures, the stable form of oxygen is the diatomic gas, but rhombic sulphur consists of eight membered rings held together in a crystal by van der Waals forces:

$$S\,(\text{rhombic}) \rightarrow \tfrac{1}{8}S_8(g), \quad \Delta H^0 = +3 \cdot 1 \text{ kcal.} \tag{7.40}$$

Rhombic sulphur metals at $112 \cdot 8°$. The resulting liquid is thought to contain rings and long chains of atoms in equilibrium with one another. Their proportions and formulae vary with temperature, but enthalpies of interconversion are small because the number of S—S bonds remains nearly constant. If the liquid is heated above $160°$ and then rapidly quenched, a metastable phase called plastic sulphur is obtained. This consists of long chains of sulphur atoms. just above the boiling point $(444 \cdot 6°)$, sulphur vapour is mainly composed of S_8 molecules, but a further increase in temperature brings about dissociation of the rings to the paramagnetic, diatomic gas, for the entropy change in the process:

$$S_8(g) \rightarrow 4S_2(g), \quad \Delta H^0 = +98 \cdot 3 \text{ kcal,} \tag{7.41}$$

is very favourable. The standard enthalpy change of the corresponding reaction in oxygen chemistry may be estimated from the data in table 7.2 when a value of -205 kcal/mole is obtained. Thus O—O rings or chains are very unstable to the diatomic molecule. Like the low dissociation energy of fluorine, the weakness of the O—O bond may be due to repulsions between the non-bonding electrons across the short internuclear distances which prevail.

Not surprisingly, the tendency of sulphur to catenate in its elemental forms is reflected in the formation of a number of compounds which have no oxygen analogues. These compounds, however, are not very stable with respect to sulphur abstraction from the chain. The hydrogen persulphides have the formula H_2S_n where n may be as high as eight. In the gas phase, there is no change in the number of S—S bonds in the process

$$H_2S_n \rightarrow H_2S + \frac{n-1}{8}S_8 \tag{7.42}$$

and the reaction would be thermally neutral if bond energies were independent of their environment. In practice, the values of the enthalpy of condensation of the liquid polysulphide and the heat of formation of rhombic sulphur from S_8 rings are such that the reaction is endothermic at room temperature, but these perturbations are small:

$$H_2S_4(l) \rightarrow H_2S(g) + 3S(s), \quad \Delta H^0 = 2 \cdot 9 \text{ kcal} \tag{7.43}$$

$$H_2S_6(l) \rightarrow H_2S(g) + 5S(s), \quad \Delta H^0 = 3 \cdot 9 \text{ kcal}. \tag{7.44}$$

Owing to the formation of gaseous H_2S, the entropy change is positive, and consequently quite mild heating is usually sufficient to initiate reactions of type [7.42]. Similarly, both the standard free energy and enthalpy changes for the deposition of sulphur from the chain of a polysulphide anion are close to zero. For example, for the reaction,

$$S_4^{2-}(aq) + S(c) \rightarrow S_5^{2-}(aq), \tag{7.45}$$

$$\Delta H^0 = -0 \cdot 4 \text{ kcal} \quad \text{and} \quad \Delta G^0 = -0 \cdot 8 \text{ kcal}$$

and in aqueous solution, the Na_2S/S equilibrium diagram shows that all the polysulphides from Na_2S to Na_2S_5 are present. By contrast, the relative strengths of the O—O single and double bonds ensures that the abstraction of oxygen from O_n chains in the gas phase is a highly exothermic process, and both the ΔH^0 and ΔS^0 terms favour decomposition. This tendency is unaffected by relatively small heats of condensation or solution; H_2O_2 is the only well-characterized hydrogen polyoxide known, and even this slowly decomposes to water in aqueous solution.

Apart from the diatomic molecule, the only other well characterized oxide of oxygen is ozone. This molecule has an O—O—O bond angle of $116 \cdot 8°$ and two equal O—O bond lengths of $1 \cdot 278$ Å. From the standard enthalpy of the reaction,

$$3O_2(g) \rightarrow 2O_3(g), \quad \Delta H^0 = +68 \cdot 2 \text{ kcal/mole}, \tag{7.46}$$

the oxygen–oxygen bond energy term is 72 kcal/mole. Thus both the lengths and energies of the bonds lie between the figures of $1 \cdot 21$ and $1 \cdot 49$ Å, and of 119 and 34 kcal/mole for the corresponding properties of the bonds in O_2 and H_2O_2. These facts imply a bond order between one and two, while the bond angle suggests a valence bond description of the molecule in terms of sp^2 hybridization of the central oxygen. Qualitatively therefore, an interpretation of the bonding with structure V is consistent with the experimental facts.

V

TABLE 7.10   *Bond lengths and bond energy terms in the oxides of sulphur*

	SO	SO_2	SO_3
Bond length (Å)	1·49	1·43	1·43
Bond energy term (kcal/mole)	125	128	113

The O—S—O angle is SO_2 at 119·5°, is very close to the value described by sp^2 hybridization. However, in contrast to the relative values of the corresponding properties of their oxygen analogues, both the bond lengths and bond energies in SO_2 and SO indicate a higher bond order in the triatomic molecule (see table 7.10). This suggests that a structural interpretation of the sulphur dioxide molecule based upon sp^2 hybridization, and similar to that given for ozone, would be inadequate. An additional contribution to the bonding has been proposed by Moffitt (1950). He suggested that in SO_2 and gaseous SO_3, $(p_\pi-d_\pi)$ bonding occurs between oxygen and sulphur. Thus a more appropriate, although still approximate description of these structures is provided by VI and VII.

VI   VII

The inverse correlation between bond length and bond energy term breaks down in the oxides of sulphur, but the differences between the values for the three molecules are rather small.

Although the value of B(S—O), where both sulphur and oxygen are divalent, is unknown a figure of 65 kcal/mole may be calculated from (7.10) and the data in table 7.6. This yields

$$\Delta H_f^0 [SO] = -4 \text{ kcal/mole}$$

for a chain polymer $(SO)_n$ which consists of alternate, divalent sulphur and oxygen atoms. Weak van der Waals forces between individual chains would presumably result in only small heats of condensation, and the estimated value, $\Delta H^0 = -63$ kcal, for the reaction,

$$2SO \rightarrow S(s) + SO_2(g), \qquad [7.47]$$

suggests that the unknown compound is thermodynamically unstable with respect to disproportionation. $(SO)_n$ rings, of course, should be equally unstable. In the presence of an excess of the dioxides, small concentrations of the diatomic molecule, SO, exists in equilibrium with S_2 and SO_2 at 1550°. This contains a sulphur–oxygen multiple bond.

With fluorine and chlorine, sulphur attains high oxidation states in the compounds SF_6 and SCl_4. Sulphur dichloride is well characterized but the dibromide and diiodide are unknown while the existence of the difluoride is very doubtful. From (7.10) and the data in table 7.6, we calculate 85, 56 and 48 kcal/mole for $B(S\text{—}F)$, $B(S\text{—}Br)$ and $B(S\text{—}I)$ respectively. These values yield -66, $+8$ and $+22$ kcal/mole for the enthalpies of formation of SF_2, SBr_2 and SI_2 in the gas phase. As the heats of condensation are likely to be small and the entropies of decomposition into the elements favourable, it is likely that SBr_2 and SI_2 will be thermodynamically unstable with respect to sulphur and the appropriate halogen. The value of $\Delta H_f^0 [SF_2(g)]$ estimated above implies thermodynamic instability with respect to disproportionation:

$$3SF_2(g) \rightarrow 2S(c) + SF_6(g), \quad \Delta H^0 = -94 \text{ kcal.} \qquad [7.48]$$

As sulphur hexafluoride is thermodynamically unstable with respect to hydrolysis:

$$SF_6(g) + 4H_2O(l) \rightarrow H_2SO_4(aq) + 6HF(aq),$$
$$\Delta G^0 = -110 \text{ kcal,} \qquad [7.49]$$

the absence of any reaction with water must be due to kinetic factors. This observation is interesting in view of the availability of vacant $3d$ orbitals for coordination with an attacking nucleophile.

7.13. The halogens. A number of aspects of halogen chemistry have already been discussed in other chapters of this book where attempts were made to interpret certain reactions of the elements and their compounds by means of an ionic model. The chemistry of more volatile halogen compounds cannot be treated in this way

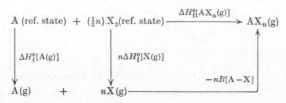

Fig. 7.6. Thermodynamic cycle for the formation of a gaseous halide, AX_n, from its elements.

because a relation between internuclear distance and the energies of processes such as,

$$SF_6(g) \rightarrow S^{6+}(g) + 6F^-(g) \qquad\qquad [7.50]$$

and

$$MoCl_5(g) \rightarrow Mo^{5+}(g) + 5Cl^-(g) \qquad\qquad [7.51]$$

has not yet been established. An alternative approach, based upon the bond model, is also rather unsatisfactory because there is no adequate interpretation of the variations in bond energy terms. Nevertheless, certain helpful correlations may be observed.

We shall consider first the formation of a halide AX_n from the elements A and X_2 in their standard reference states. By means of the equation derived from the cycle in fig. 7.6, the enthalpy of formation of the gaseous compound may be expressed in terms of the heats of atomization of the elements and $B(A-X)$:

$$\Delta H_f^0[AX_n(g)] = \Delta H_f^0[A(g)] + n\Delta H_f^0[X(g)] - nB(A-X). \quad (7.14)$$

In comparing the stabilities of the four halides in this oxidation state with respect to the decomposition,

$$AX_n(\text{usual state}) \rightarrow A(\text{ref. state}) + \frac{n}{2}X_2(\text{ref. state}) \qquad [7.52]$$

we shall assume that, at 25°, the enthalpy change associated with any condensation of the gaseous molecular halide, AX_n, is small, and that, from halogen to halogen, the variations in the $T\Delta S^0$ terms in [7.52] are small compared with those in ΔH^0. As such assumptions are usually justified for the volatile halides that we shall consider, the changes in the standard enthalpy of the reaction,

$$AX_n(g) \rightarrow A(\text{ref. state}) + \frac{n}{2} X_2(\text{ref. state}) \qquad [7.53]$$

may be regarded as a safe guide to those in the standard free energy of [7.52].

TABLE 7.11 *Enthalpies of formation of halogen atoms in the gas phase at 25° (kcal/mole)*

	F	Cl	Br	I
$\Delta H_f^0[X_2(g)]$	0	0	7·4	14·9
Dissociation enthalpy	37·8	58·2	46·1	36·2
$2\Delta H_f^0[X(g)]$	37·8	58·2	53·5	51·1

In table 7.11, the enthalpies of atomization of the halogens are expressed as the sum of the vaporization energies, where applicable, and the dissociation enthalpies. In the case of chlorine, bromine and iodine, it is apparent that considerable variations in the latter quantity are partially offset by the vaporization energies of liquid bromine and solid iodine, and that the resulting differences in the enthalpies of atomization are small. The dissociation enthalpy of fluorine however, is considerably less than the enthalpies of atomization of the other three halogens.

The data in table 7.2 show that in nearly every case, the order of bond energies in any halide, AX_n, is F > Cl > Br > I, and that this variation is considerable. Since the differences in $\Delta H_f^0[X(g)]$ are small for Cl, Br and I, substitution of this sequence in (7.14) implies that the stability order of the three halides is Cl > Br > I. For the fluorides, the low dissociation energy of the halogen only enhances the relative stability of the compounds, so the final order of stability of the halide MX_n, with respect to its constituent elements, is F > Cl > Br > I.

Thus NF_3 is an exothermic compound while NCl_3 detonates with the formation of its elements at the slightest provocation. NBr_3 and NI_3 have not been prepared. The endothermic compound OCl_2 is unstable with respect to its elements at 25° and explodes on heating or sparking. OF_2 is thermodynamically stable to this reaction at room temperature, and dissociation into fluorine and oxygen only begins at 250°. OBr_2 begins to decompose above $-50°$ and OI_2 is unknown. Again, the dissociation of hydrogen iodide on mild heating contrasts with the explosive formation of HF and HCl from their elements at room temperature and the combination of hydrogen and bromine to give hydrogen bromide in the presence of a platinum catalyst at 200°. The lower thermal stability of iodides is frequently used for the preparation of pure elements. By the pyrolysis of the iodides on hot wires at about 1000°, very pure silicon, for use in semi-conductor devices, and boron may be obtained. The same technique, due to Van Arkel and De Boer, may be used for the preparation of high purity metals like titanium, zirconium, thorium and uranium from their tetraiodides.

That a lighter halogen will often displace a heavier from a compound MX_n is well known. Thermodynamically, this is a necessary corollary of the observed variation in $\Delta H_f^0 [MX_n]$ and the relative insignificance of the entropy contribution to the order of free energies of formation.

The relative capacities of the four halogens to bring out high oxidation states, however, is a question which involves more than the order of the values of $\Delta G_f^0 [MX_n]$. On dissociation, a halide usually yields the halide of a lower oxidation state rather than the elements of which it is composed. Thus on heating, phosphorus pentabromide and ruthenium hexafluoride decompose according to the reactions,

$$PBr_5 \rightarrow PBr_3 + Br_2 \qquad\qquad [7.54]$$

and

$$RuF_6 \rightarrow RuF_5 + \tfrac{1}{2}F_2. \qquad\qquad [7.55]$$

Suppose that $B(A\text{---}X)$ in a halide AX_q is $B_q(A\text{---}X)$, then if $n > m$, for the process

$$AX_n(g) \rightarrow AX_m(g) + \frac{n-m}{2} X_2(\text{ref. state}) \qquad\qquad [7.56]$$

we have,

$$\Delta H^0 = nB_n(A\!-\!X) - mB_m(A\!-\!X) - \frac{n-m}{2}\Delta H_f^0[X(g)]. \qquad (7.15)$$

As we saw earlier in this section, the last term in this equation suggests that the order of stability of AX_n should be F > Cl < Br < I but the differences between the last three are rather small. It is impossible, however, to say anything very definite about the variations in $B_q(A\!-\!X)$ as q is increased. Often it falls sharply (ClF, 60; ClF$_3$, 41; PCl$_3$, 79; PCl$_5$, 63), sometimes it stays almost constant (XeF$_2$, 31; XeF$_4$, 31; XeF$_6$, 30) and sometimes irregularities are observed (IF, 47; IF$_5$, 63; IF$_7$, 55). However, unless the differences between $B_m(A\!-\!X)$ and $B_n(A\!-\!X)$ are very dissimilar for the four halogens, because n exceeds m and the order of values of $B_q(A\!-\!X)$ for any value of q is F > Cl > Br > I, the same sequence will be observed in $[nB_n(A\!-\!X) - mB_m(A\!-\!X)]$. This is nearly always the case. When this effect is imposed upon the variations in the values of $\frac{n-m}{2}\Delta H_f^0[X(g)]$, the stability order for AX_n becomes F > Cl > Br > I. Changes in the entropies of decomposition and in the differences in any heats of condensation of the two halides have again been neglected.

Among the many fluorides which contain oxidation states of elements higher than those found in combination with any other halogen are AsF$_5$, BiF$_5$, BrF$_5$, CrF$_6$, IF$_7$, IrF$_6$, OsF$_6$, RhF$_6$, RuF$_6$, SeF$_6$, SF$_6$, VF$_5$ and XeF$_6$. The highest oxidation states formed with each halogen by the elements rhenium and iodine are found in the compounds ReF$_7$, ReCl$_6$, ReBr$_5$ and ReI$_4$ and IF$_7$, ICl$_3$, IBr and I$_2$.

The order of halide stability for the more volatile compounds discussed in this section is identical with that obtained by the application of the ionic model to the solid halides in §2.8. For a solid halide MX_n, from (2.2),

$$\Delta H_f^0[MX_n(s)] = \Delta H_f^0[M^{n+}(g)] + n\Delta H_f^0[X^-(g)]$$
$$- U[MX_n] - (n+1)RT. \qquad (7.16)$$

The data in table 2.10 show that the term $-\Delta H_f^0[X^-(g)]$, like the lattice energies, varies in the order F > Cl > Br > I. This is therefore the sequence in the stabilities of the four halides with respect to the elements.

8

Thus both platinum and palladium dichlorides can be made by direct combination of the elements at about 500°, a temperature above that at which the di-iodides decompose to the metals. The stabilities of the halides with respect to others in lower oxidation states was discussed in §2.8.

The bond energy treatment of halide stability presented in this section compares unfavourably with the interpretation by the ionic model. It is in fact no more than a restatement of the problem because bond energies cannot be theoretically understood in the way that lattice energies can. That the order of stabilities is the same in both types of halide is a tribute to the way in which the conclusions of the ionic model are often applicable, even when the compounds under consideration are far from ionic.

The electrode potentials of the halogens, $E^0[\frac{1}{2}X_2/X^-]$, (see table 4.2) were discussed on page 91. Because the values of $\Delta H_f^0[X^-(g)]$, and consequently $\Delta G_f^0[X^-(g)]$, are not very different, the substantial variations in the hydration energies of the four anions determines the potential sequence.

In acid and neutral solution, the electrode potentials of the couple,

$$O_2 + 4H^+ + 4e \rightarrow 2H_2O \qquad\qquad [7.57]$$

are 1·23 and 0·81 V respectively.

According to these values, appreciable pressures of chlorine are capable of oxidizing water under neutral or acid conditions. The reaction is slow however, and disproportionation is the initial result when water is saturated with the gas:

$$Cl_2 + H_2O \rightarrow HClO + HCl, \quad \Delta G^0 = 6\cdot2 \text{ kcal.} \qquad [7.58]$$

The equilibrium position lies to the left in the resulting solution, but the hypochlorous acid decomposes slowly:

$$HClO \rightarrow HCl + \tfrac{1}{2}O_2, \quad \Delta G^0 = -12\cdot3 \text{ kcal.} \qquad [7.59]$$

Thus although it is not fast, the oxidation of water is the overall result.

Disproportionation into HBrO and HBr occurs in bromine water but the equilibrium position lies even further to the left than in [7.58]. One slow decomposition of hypobromous acid proceeds according to the equation:

$$2HBrO \rightarrow Br_2 + H_2O + \tfrac{1}{2}O_2 \qquad\qquad [7.60]$$

so oxidation of water again occurs. Creation of fresh HBrO and HBr by the disproportionation of bromine raises the acidity however, and eventually stops the reaction for, as the potentials show, the oxidation of water by bromine does not proceed at unit activity of hydrogen ions. Indeed solutions of hydrobromic acid slowly redden due to aerial oxidation of bromide. Aqueous iodides react with atmospheric oxygen in both acid and neutral conditions, although in the second case the reaction is extremely slow.

Fluorine reacts very quickly with water producing ozone as well as oxygen. The potential required for the direct oxidation of water to ozone:

$$O_3 + 6H^+ + 6e \rightarrow 3H_2O, \qquad\qquad [7.61]$$

is 1·51 V. No intermediate has yet been observed in the fluorine oxidation, and hypofluorous acid, along with other oxyacids of fluorine, is unknown. From table 7.2, $\Delta H_f^0 [\text{HOF(g)}] = -31$ kcal/mole, and for the reaction,

$$\text{HOF(g)} \rightarrow \text{HF(g)} + \tfrac{1}{2}O_2, \quad \Delta H^0 = -34 \text{ kcal.} \qquad [7.62]$$

If the heat of solution were the same as that of HOCl, then ΔH^0 for the reaction,

$$\text{HOF(aq)} \rightarrow \text{HF(aq)} + \tfrac{1}{2}O_2(g) \qquad\qquad [7.63]$$

would be -39 kcal. As the entropy change in both the gas phase and solution decompositions would be favourable, the compound is thermodynamically unstable with respect to hydrogen fluoride and oxygen.

The great reactivity of the fluorine molecule is due to its ability to form fluorides that are thermodynamically stable with respect to their constituent elements by means of reactions that are fast. The thermodynamic stability of these compounds arises from the strong bonds that fluorine forms with other elements and the weak bond that it forms with itself. This last property probably accounts for the speed of reaction, for the production of fluorine atoms in a rate determining step is a relatively easy matter.

It is worth noting that it is tautologous to attribute the reactivity of fluorine to its high electronegativity. According to the traditional Pauling definition, high electronegativities arise automatically when the single bonds that an element forms with other atoms are much stronger than those that it forms with itself. It is as correct to say that fluorine is very electronegative because it is very reactive.

Another important electronegativity scale is due to Mulliken who defined the property as the mean of the sum of the ionization potential and electron affinity of the atom. Since the electron affinities of the halogens are quite similar, the electronegativity sequence F > Cl > Br > I is determined by the variations in the ionization potential a measure of the difficulty of electron release.

References and suggestions for further reading

Benson, S. W. (1965). *J. Chem. Educ.*, **42**, 502. A very instructive article which deals with bond dissociation energies.

Coates, G. E. & Sutton, L. E. (1948). *J. Chem. Soc.* 1187.

Cotton, F. A. & Wilkinson, G. (1966). *Advanced Inorganic Chemistry, a Comprehensive Text*, 2nd edition. New York: Interscience.

Cottrell, T. L. (1958). *The Strengths of Chemical Bonds*, 2nd edition. London: Butterworths. A thorough treatment of thermodynamic measures of bond strength with valuable experimental details.

Hartley, S. B., Holmes, W. S., Jacques, J. K., Mole, M. F. & McCoubrey, J. C. (1963). *Quart. Revs*, 204.

Ives, D. J. G. (1960). *Principles of the Extraction of Metals*, R.I.C. Monographs for Teachers, pp. 12–33. Includes a discussion of the value of carbon in metal extraction.

Moffitt, W. (1950). *Proc. Roy. Soc.* A, **200**, 409.

Wells, A. F. (1962). *Structural Inorganic Chemistry*, 3rd edition. Oxford University Press.

Appendix 1

Symbols, standard states, definitions, conventions and sources of data

Some symbols used in the text

a	activity	log	logarithm to base 10
B	bond energy term	M	Madelung constant
B_I	intrinsic bond energy	m	molality or number of pairs of parallel spins
C_p	molal heat capacity at constant pressure	N_0	Avogadro number
C_v	molal heat capacity at constant volume	p	pressure
		P_n	nth ionization enthalpy
D	bond dissociation energy	q	heat absorbed
E	internal energy, e.m.f. of cell or activation energy	R	gas constant
		r	radius of ion
E_a	electron affinity	r_0	internuclear distance
e	charge on an electron	S	entropy
F	Faraday constant	T	temperature (°K)
f	fugacity	U_T	lattice energy at T °K
G	free energy	V	volume
H	enthalpy	z_+, z_-	charge on cation, anion
I_n	nth ionization potential	Δ	ligand field splitting
$\sum_l^h I_n$ $\quad I_{l+1} + I_{l+2} + \dots I_h$		ϵ	dielectric constant
		γ	activity coefficient
K	equilibrium constant	ν	no. of ions in one molecule
k	velocity constant		
L_v	latent heat of vaporization	ΔG_h	free energy of hydration
		ΔH_h	enthalpy of hydration
ln	logarithm to base e		

Temperature subscripts to thermodynamic properties are in degrees Kelvin. With this exception, temperatures are in degrees centigrade unless otherwise stated.

Standard states

A superscript zero against a thermodynamic property of a reaction indicates that the reactants and products are in their standard states.

(a) For a pure solid or liquid, the standard state is the substance in the condensed phase under a pressure of one atmosphere.

(b) For a gas, the standard state is the hypothetical ideal gas state at one atmosphere pressure. In this state, the molal enthalpy is the same as that of the real gas at zero pressure.

(c) For a solute in an aqueous solution, the standard state is the hypothetical ideal solution of unit molality. In this state, the partial molal enthalpy of the solute is the same as in the infinitely dilute real solution.

Definitions of some quantities used in the text

(a) The standard free energy of formation of a substance, ΔG_f^0, is the free energy change when one mole of the substance in its standard state is formed at a specified temperature from its elements in their standard reference states.

The standard enthalpy and entropy of formation, ΔH_f^0 and ΔS_f^0, are the corresponding enthalpy and entropy changes.

At 25°, the National Bureau of Standards takes the standard reference state for every element except phosphorus to be the standard state of that form of the element that is thermodynamically stable with respect to all other forms at 25°. For phosphorus, the standard reference state is the crystalline white form at one atmosphere pressure.

(b) The standard molal entropy of a substance, S^0, is its molal entropy in the standard state at a specified temperature, the entropy of a perfect crystal of that substance being defined to approach zero as the absolute temperature approaches zero.

(c) The latent heat or standard enthalpy of vaporization of a substance, L_v, is the enthalpy change associated with the transfer at a specified temperature, of one mole of the substance from the condensed phase to a specified form in the gas phase, condensed and gaseous forms being in their standard states.

(d) Bond dissociation energies, bond energy terms and intrinsic bond energies: see §7.2–7.4.

(e) The ionization potential, I, of a species, A, is defined here as the molal internal energy change of the reaction,

$$A(g) \rightarrow A^+(g) + e^-(g) \tag{A1.1}$$

at 0 °K, reactants and products being in their standard states.

(f) The electron affinity, E_a, of a species, A, is defined here as the molal internal energy change of the reaction,

$$A(g) + e^-(g) \rightarrow A^-(g) \tag{A1.2}$$

at 0 °K, reactants and products being in their standard states.

(g) The lattice energy, U_T, of a solid compound is the internal energy change when one mole of the compound at a pressure of one atmosphere is converted, at T °K into defined, gaseous ions which are infinitely removed from one another. Only at 0 °K are these ions stationary.

Conventions

(i) As values of ΔG_f^0 and ΔH_f^0 refer to formation from the elements in their standard reference states, values of ΔG_f^0 and ΔH_f^0 for the elements in their standard reference states are zero at all temperatures by convention.

(ii) The molal entropy of a perfect crystal of a substance is defined to approach zero as the absolute temperature approaches zero.

(iii) Values of ΔH_f^0 for gaseous ions are calculated by using a convention on which ΔH_f^0 for the gaseous electron is zero at all temperatures, e.g. $\Delta H_f^0 [Na^+(g)]$ at any temperature is the standard enthalpy of the reaction:

$$Na(s) \rightarrow Na^+(g) + e^-(g) \tag{A1.3}$$

at that temperature. The calculation of the enthalpies of formation of gaseous ions using ionization potentials and electron affinities is described in appendix 2.

(iv) Conventions regarding the thermodynamic properties of aquated ions, the e.m.f. of cells and electrode potentials are treated together in §4.1–4.4.

Sources and thermodynamic data

Thermodynamic data were taken principally from the U.S. National Bureau of Standards Technical Notes 270–1 and 270–2,

U.S. Government Printing Office, Washington, 1965 and 1966, and, when not available in these two publications, from the U.S. National Bureau of Standards Circular 500, U.S. Government Printing Office, Washington, 1952.

Some valuable revisions of figures given in this last source, and a considerable amount of new data were found in K. S. Pitzer and L. Brewer's revision of G. N. Lewis and M. Randall's *Thermodynamics*, McGraw Hill, N.Y., 1961. Useful supplementary sources were W. M. Latimer's *The Oxidation States of the Elements and their Potentials in Aqueous Solution*, 2nd edition, Prentice Hall, N.Y., 1952 and K. K. Kelley and E. G. King's *Entropies of the Elements and Inorganic Compounds*, Bulletin 592 of the U.S. Bureau of Mines, U.S. Government Printing Office, Washington, 1961. Integrated values of C_p^0 for halides, MX_n, between 0 °K and 298 °K were taken from D. Cubicciotti, *J. Chem. Phys.*, 1959, **31**, 1646, and L. Brewer, G. R. Somayajulu and E. Brackett, *Chem. Rev.*, 1963, **63**, 111.

Appendix 2

Derivation of some terms in fig. 2.3

It is assumed that only the ground states of $M(g)$, $M^{n+}(g)$, $X(g)$ and $X^-(g)$ are appreciably populated at the temperature T.

(*a*) For the reaction,

$$MX_n(s) \rightarrow M^{n+}(g) + nX^-(g), \qquad\qquad [A2.1]$$

$$\Delta E^0_{298} = U_{298}[MX_n], \qquad\qquad (A2.1)$$

because the internal energy of ideal gases at constant temperature is the same whether they are infinitely attenuated or at one atmosphere pressure. Now,

$$\Delta H^0_{298} = \Delta E^0_{298} + p\Delta V \qquad\qquad (A2.2)$$

$$= U_{298}[MX_n] + (n+1)RT, \qquad\qquad (A2.3)$$

neglecting the volume of the solid.

(*b*) For the reaction,

$$M(g) \rightarrow M^{n+}(g) + ne^-(g) \qquad\qquad [A2.2]$$

$$\Delta E^0_0 = \sum_0^n I_n = \Delta H^0_0. \qquad\qquad (A2.4)$$

According to Kirchoff's law,

$$\left(\frac{\partial \Delta H^0}{\partial T}\right)_p = \Delta C^0_p \qquad\qquad (A2.5)$$

$$\therefore \Delta H^0_T = \Delta H^0_0 + \int_0^T \Delta C^0_p \, dT. \qquad\qquad (A2.6)$$

For ideal monatomic gases, $C^0_p = \frac{5}{2}R$, so

$$\Delta H^0_T = \sum_0^n I_n + \frac{5}{2}nRT. \qquad\qquad (A2.7)$$

(c) For the reaction,

$$X(g) + e^-(g) \to X^-(g), \qquad\qquad\qquad [A\,2.3]$$

$$\Delta E_0^0 = -E_a = \Delta H_0^0, \qquad\qquad\qquad (A\,2.8)$$

and from (A 2.6),

$$\Delta H_T^0 = -E_a - \tfrac{5}{2}RT. \qquad\qquad\qquad (A\,2.9)$$

(d) Evaluation of $U_T[\mathrm{MX}_n]$ from $U_0[\mathrm{MX}_n]$ may be carried out as follows: ignoring the pV term for the solid, for [A 2.1],

$$\Delta H_0^0 = \Delta E_0^0. \qquad\qquad\qquad (A\,2.10)$$

Using (A 2.3), (A 2.6) and $C_p^0 = \tfrac{5}{2}R$ for a monatomic gas, we obtain,

$$U_T[\mathrm{MX}_n] + (n+1)RT = U_0[\mathrm{MX}_n] + \int_0^T \{\tfrac{5}{2}R(n+1)$$

$$- C_p^0[\mathrm{MX}_n]\}\mathrm{d}T. \quad (A\,2.11)$$

$$\therefore\ U_T[\mathrm{MX}_n] - U_0[\mathrm{MX}_n] = \tfrac{3}{2}RT(n+1)$$

$$- \int_0^T C_p^0[\mathrm{MX}_n]\mathrm{d}T. \quad (A\,2.12)$$

If the specific heat–temperature function of the compound MX_n is known from 0–T °K, the integration may be performed and $(U_T[\mathrm{MX}_n] - U_0[\mathrm{MX}_n])$ determined.

Appendix 3

Determination of n in (2.7) from the isothermal compressibility

$$\mathrm{d}E = T\mathrm{d}S - p\mathrm{d}V. \tag{A 3.1}$$

$$\therefore \text{ At } 0 \text{ °K } \mathrm{d}E/\mathrm{d}V = -p. \tag{A 3.2}$$

The isothermal compressibility of the crystal, β, is defined by the equation,

$$\beta = -1/V(\mathrm{d}V/\mathrm{d}p)_T. \tag{A 3.3}$$

As any increase in E_0 for the solid decreases U_0 by an equal amount $\mathrm{d}E$ may be replaced by $-\mathrm{d}U_0$ at 0 °K. Then,

$$\mathrm{d}^2U_0/\mathrm{d}V^2 = \mathrm{d}p/\mathrm{d}V = -1/V\beta. \tag{A 3.4}$$

V, the volume containing one gram molecule, is related to the internuclear distance, r, by the equation,

$$V = kr^3, \tag{A 3.5}$$

where k is a constant for a particular type of structure.

$$\mathrm{d}U_0/\mathrm{d}V = \mathrm{d}r/\mathrm{d}V . \mathrm{d}U_0/\mathrm{d}r = \frac{1}{3kr^2}.\mathrm{d}U_0/\mathrm{d}r, \tag{A 3.6}$$

$$\mathrm{d}^2U_0/\mathrm{d}V^2 = \mathrm{d}r/\mathrm{d}V . \mathrm{d}[\mathrm{d}U_0/\mathrm{d}V]/\mathrm{d}r, \tag{A 3.7}$$

$$= \frac{1}{3kr^2}\left[\frac{1}{3kr^2}.\mathrm{d}^2U_0/\mathrm{d}r^2 - \frac{2}{3kr^3}.\mathrm{d}U_0/\mathrm{d}r\right]. \tag{A 3.8}$$

$$\therefore 1/\beta = -V\left[\frac{1}{9k^2r^4}.\mathrm{d}^2U_0/\mathrm{d}r^2 - \frac{2}{9k^2r^5}.\mathrm{d}U_0/\mathrm{d}r\right], \tag{A 3.9}$$

$$= 1/(9V)[2r.\mathrm{d}U_0/\mathrm{d}r - r^2.\mathrm{d}^2U_0/\mathrm{d}r^2]. \tag{A 3.10}$$

Now from (2.5) and (2.6), putting $N_0 M z_+ z_- e^2 = A$,

$$dU_0/dr = -A/r^2 + A r_0^{n-1}/r^{n+1}. \tag{A 3.11}$$

$$\therefore \ d^2U_0/dr^2 = 2A/r^3 - A(n+1) r_0^{n-1}/r^{n+2}. \tag{A 3.12}$$

Substituting in (A 3.10) and putting $r = r_0$,

$$1/\beta = A(n-1)/9 V r_0, \tag{A 3.13}$$

$$\therefore \ n = 1 + 9 V r_0/\beta N_0 M z_+ z_- e^2 \tag{A 3.14}$$

V and r_0 may be determined by X-ray crystallography while β may be obtained by measuring change of volume with pressure and using (A 3.3). If these quantities are extrapolated to 0 °K, (A 3.14) then yields n which usually lies between 5 and 12.

Appendix 4

Ionization potentials of the elements (kcal/mole)

Element	1	2	3	4	5
H	314	—	—	—	—
He	567	1255	—	—	—
Li	124	1744	2823	—	—
Be	215	420	3548	5019	—
B	191	580	874	5980	7844
C	260	562	1104	1487	9040
N	335	682	1094	1786	2257
O	314	810	1266	1785	2626
F	402	807	1445	2010	2634
Ne	497	947	1464	2237	2913
Na	118	1091	1652	2280	3191
Mg	176	347	1848	2520	3257
Al	138	434	656	2766	3546
Si	188	377	772	1041	3845
P	242	455	695	1184	1499
S	239	540	807	1091	1672
Cl	300	549	920	1234	1564
Ar	363	637	943	1379	1730
K	100	734	1061	1404	1905
Ca	141	274	1181	1545	1946
Sc	151	295	571	1704	2122
Ti	157	313	633	997	2301
V	155	338	676	1107	1504
Cr	156	380	714	1144	1683
Mn	171	361	777	—	1753
Fe	182	373	707	—	—
Co	181	393	772	—	—
Ni	176	418	812	—	—
Cu	178	468	849	—	—
Zn	217	414	916	—	—
Ga	138	473	708	1481	—
Ge	182	367	789	1054	2154
As	226	430	654	1155	1444
Se	225	496	738	989	1575
Br	273	498	828	1091	1377
Kr	323	566	851	—	—
Rb	96	634	922	—	—
Sr	131	254	—	1314	—
Y	147	282	473	—	1776

Ionization potentials of the elements (kcal/mole)

Element	1	2	3	4	5
Zr	158	303	530	792	—
Nb	159	330	577	883	1153
Mo	164	372	626	1070	1411
Tc	168	352	—	—	—
Ru	170	387	656	—	—
Rh	172	417	716	—	—
Pd	192	448	921	—	—
Ag	175	495	803	—	—
Cd	207	390	864	—	—
In	133	435	646	1255	—
Sn	169	337	703	939	1667
Sb	199	381	583	1017	1291
Te	208	429	715	876	1384
I	241	440	—	—	—
Xe	280	489	740	—	—
Cs	90	579	798	—	—
Ba	120	231	—	—	—
La	129	255	442	—	—
Ce	129	250	454	847	—
Pr	126	243	535	—	—
Nd	127	247	—	—	—
Pm	—	—	—	—	—
Sm	129	255	—	—	—
Eu	131	259	—	—	—
Gd	142	279	—	—	—
Tb	138	266	—	—	—
Dy	137	269	—	—	—
Ho	139	272	—	—	—
Er	140	275	—	—	—
Tm	134	278	—	—	—
Yb	143	278	586	—	—
Lu	142	323	—	—	—
Hf	—	344	—	—	—
Ta	182	374	—	—	—
W	184	408	—	—	—
Re	181	383	—	—	—
Os	201	392	—	—	—
Ir	208	—	—	—	—
Pt	208	428	—	—	—
Au	213	473	—	—	—
Hg	241	432	789	—	—
Tl	141	471	687	1169	—
Pb	171	347	736	976	1587
Bi	168	385	589	1045	1291
Po	194	—	—	—	—
At	—	—	—	—	—
Rn	248	—	—	—	—
Fr	—	—	—	—	—

Element	1	2	3	4	5
Ra	122	234	—	—	—
Ac	—	279	—	—	—
Th	160	—	680	—	—
Pa	—	—	—	—	—
U	140	—	—	—	—
Np	—	—	—	—	—
Pu	134	—	—	—	—
Am	138	—	—	—	—

Figures are mainly from C. E. Moore, *Atomic Energy Levels*, U.S. National Bureau of Standards Circular 467, vols. I, II and III, U.S. Government Printing Office, Washington, 1949, 1952 and 1958, *Appl. Optics*, 1963, **2**, 665, and J. Sugar and J. Reader, *J. Opt. Soc. Am.*, 1965, **55**, 1286. Values in the appendices of vols. II and III of Circular 467 were assumed to supercede those in the text. The conversion factor 1 eV = 23·061 kcal/mole was employed and the resultant figures were rounded to the nearest kcal/mole. In many cases the uncertainties are considerably greater than this.

Appendix 5

Selected thermodynamic data at 25°

Compound	State	ΔH_f^0 (kcal/mole)	ΔG_f^0 (kcal/mole)	S^0 (cal/deg.mole)
Ag	c	0	0	10·21
Ag	g	68·1	58·9	41·32
Ag$^+$	g	244·2	—	—
Ag$^+$	aq	25·31	18·43	17·67
AgF	c	−48·5	−44·2	20
AgCl	c	−30·36	−26·22	22·97
AgBr	c	−23·8	−22·93	25·60
AgI	c	−14·9	−15·85	27·3
As	$\alpha-$c	0	0	8·4
As$_4$	g	34·4	22·1	75
As$_4$O$_6$	g	−289·0	−262·4	91
AsH$_3$	g	15·88	16·47	53·22
AsF$_3$	l	−228·55	−217·29	43·31
AsF$_3$	g	−220·04	−216·46	69·07
AsCl$_3$	g	−61·80	−58·77	78·17
AsBr$_3$	g	−31	−28	86·94
B	c	0	0	1·40
B	g	134·5	124·0	36·65
BO	g	6	−1	48·62
B$_2$O$_3$	c	−304·20	−285·30	12·90
B$_2$H$_6$	g	8·5	20·7	55·45
BF$_3$	g	−271·75	−267·77	60·71
B$_2$F$_4$	g	−344·2	−337·1	75·8
BCl$_3$	g	−96·50	−92·91	69·31
BBr$_3$	g	−49·15	−55·56	77·47
BI$_3$	g	17·00	4·96	83·43
Ba	c	0	0	16·0
Ba^{2+}	g	396·2	—	—
Ba^{2+}	aq	−128·67	−134·0	3
BaO	c	−133·5	−126·4	16·8
BaO$_2$	c	−150·5	—	—
Ba(OH)$_2$	c	−226·2	−204·7	—
Ba(NO$_3$)$_2$	c	−237·06	−190·0	—
BaCO$_3$	c	−291·3	−272·2	26·8

Compound	State	ΔH_f^0 (kcal/mole)	ΔG_f^0 (kcal/mole)	S^0 (cal/deg.mole)
Br_2	l	0	0	36·384
Br_2	g	7·387	0·749	58·647
Br	g	26·740	19·700	41·805
Br^-	g	$-55·9$	—	—
HBr	g	$-8·70$	$-12·77$	47·463
Br^-, HBr	aq	$-29·05$	$-24·85$	19·7
BrF	g	$-14·0$	—	—
BrCl	g	3·5	—	—
C	graphite	0	0	1·372
C	diamond	0·4533	0·6930	0·568
C	g	171·291	160·442	37·760
C_2	g	199	—	—
CO	g	$-26·416$	$-32·781$	47·219
CO_2	g	$-94·051$	$-94·258$	51·06
CO_3^{2-}	aq	$-161·84$	$-126·17$	$-13·6$
CH_4	g	$-17·88$	12·13	44·492
C_2H_6	g	$-20·24$	$-7·86$	54·85
neo–C_5H_{12}	g	$-39·67$	$-3·64$	73·23
CF_4	g	-221	-210	62·50
CCl_4	l	33·37	$-15·60$	51·72
CCl_4	g	$-24·6$	$-14·49$	74·03
CBr_4	g	19	16	85·55
Ca	c	0	0	9·95
Ca^{2+}	g	459·8	—	—
Ca^{2+}	aq	$-129·77$	132·18	13·2
CaO	c	$-151·8$	$-144·3$	9·5
CaO_2	c	$-157·5$	—	—
$Ca(OH)_2$	c	$-235·8$	$-214·33$	18·2
$Ca(NO_3)_2$	c	$-224·00$	$-177·34$	46·2
$CaCO_3$	calcite	$-288·45$	$-269·78$	22·2
$CaCO_3$	aragonite	$-288·49$	$-269·53$	21·2
Cd	α–c	0	0	12·3
Cd^{2+}	g	626·9	—	—
Cd^{2+}	aq	$-17·30$	$-18·58$	$-14·6$
CdO	c	$-60·86$	$-53·79$	13·1
CdF_2	c	$-167·4$	$-154·8$	19
CdI_2	c	$-48·4$	$-48·4$	40·2
$CdCO_3$	c	$-178·7$	$-160·2$	25·2
Cl_2	g	0	0	53·288
Cl	g	29·082	25·262	39·457
Cl^-	g	$-58·9$	—	—
HCl	g	$-22·062$	$-22·774$	44·646
Cl^-, HCl	aq	$-39·952$	$-31·372$	13·5
Cl_2O	g	19·2	23·4	63·60
HClO	g	-22	-18	56·5

Selected thermodynamic data at 25°

Compound	State	ΔH_f^0 (kcal/mole)	ΔG_f^0 (kcal/mole)	S^0 (cal/deg.mole)
HClO	aq	$-28\cdot9$	$-19\cdot1$	35
ClO_4^-	aq	$-30\cdot91$	$-2\cdot06$	$43\cdot5$
ClF	g	$-11\cdot92$	$-12\cdot28$	$52\cdot05$
ClF_3	g	$-38\cdot0$	$-28\cdot4$	$67\cdot28$
Cr	c	0	0	$5\cdot68$
Cr^{2+}	g	$634\cdot9$	—	—
Cr^{2+}	aq	-34	—	—
$CrCl_2$	c	-95	-86	$27\cdot56$
$CrCl_3$	c	-132	-115	$29\cdot4$
Cs	c	0	0	$19\cdot8$
Cs^+	g	$109\cdot9$	—	—
Cs^+	aq	$-62\cdot6$	$-70\cdot8$	$31\cdot8$
Cs_2O_2	c	$-96\cdot2$	—	—
CsO_2	c	$-62\cdot1$	—	—
CsN_3	c	$-2\cdot4$	—	—
$CsHF_2$	c	$-219\cdot5$	—	—
Cu	c	0	0	$7\cdot96$
Cu^{2+}	g	$730\cdot2$	—	—
Cu^+	aq	$17\cdot2$	$12\cdot0$	10
Cu^{2+}	aq	$15\cdot39$	$15\cdot53$	$-23\cdot6$
CuO	c	$-37\cdot1$	$-30\cdot5$	$10\cdot2$
$CuCO_3$	c	$-142\cdot2$	$-123\cdot8$	21
F_2	g	0	0	$48\cdot44$
F	g	$18\cdot88$	$14\cdot72$	$37\cdot92$
F^-	g	$-64\cdot7$	—	—
F^-	aq	$-79\cdot50$	$-66\cdot64$	$-3\cdot3$
HF	g	$-64\cdot8$	$-65\cdot3$	$41\cdot51$
HF	aq	$-76\cdot50$	$-70\cdot95$	$21\cdot2$
F_2O	g	$-5\cdot2$	$-1\cdot1$	$59\cdot1$
Fe	c	0	0	$6\cdot49$
Fe^{2+}	g	$657\cdot2$	—	—
Fe^{3+}	g	$1365\cdot3$	—	—
Fe^{2+}	aq	$-21\cdot0$	$-20\cdot30$	$-27\cdot1$
Fe^{3+}	aq	$-11\cdot4$	$-2\cdot53$	$-70\cdot1$
$FeCl_2$	c	$-81\cdot6$	$-72\cdot2$	$28\cdot19$
$FeCl_3$	c	-95	-79	$32\cdot2$
Ge	c	0	0	$7\cdot43$
Ge	g	$90\cdot0$	$80\cdot3$	$40\cdot10$
GeO_2	c	$-131\cdot7$	$-118\cdot8$	$13\cdot21$
GeH_4	g	$21\cdot7$	$27\cdot1$	$51\cdot87$
Ge_2H_6	g	$38\cdot8$	—	—

Compound	State	ΔH_f^0 (kcal/mole)	ΔG_f^0 (kcal/mole)	S^0 (cal/deg.mole)
Ge_3H_8	g	54·2	—	—
GeF_4	g	−284·4	—	—
$GeCl_4$	g	−118·5	−109·3	83·08
$GeBr_4$	g	−71·7	−76·0	94·66
GeI_4	g	−13·6	−25·4	102·5
Ge_3N_4	c	−15·1	—	—
H_2	g	0	0	31·208
H	g	52·095	48·580	27·391
H^+	g	367·161	—	—
H^+	aq	0	0	0
H^-	g	33·38	—	—
OH^-	g	−33·67	—	—
OH^-	aq	−54·970	−37·594	−2·57
H_2O	l	−68·315	−56·688	16·71
H_2O	g	−57·796	−54·635	45·10
H_2O_2	l	−44·88	−28·78	26·2
H_2O_2	g	−32·58	−25·25	55·6
Hg	l	0	0	18·5
Hg^{2+}	g	690·6	—	—
Hg^{2+}	aq	—	39·38	—
HgO	c	−21·68	−13·99	17·2
HgF_2	c	−100	—	—
$HgCl_2$	c	−53·4	—	—
$HgBr_2$	c	−40·7	—	—
HgI_2	c	−25·3	—	—
Hg_2Cl_2	c	−63·32	−50·35	46·8
Hg_2Br_2	c	−49·42	−42·71	50·9
Hg_2I_2	c	−28·91	−26·60	57·2
I_2	c	0	0	27·757
I_2	g	14·923	4·627	62·28
I	g	25·535	16·798	43·184
I^-	g	−47·0	—	—
HI	g	6·33	0·51	49·351
I^-, HI	aq	−13·19	−12·33	26·6
IF	g	−2·10	−7·56	56·42
ICl	g	4·25	−1·30	59·14
IBr	g	9·76	0·89	61·82
IF_5	g	−196·58	−179·68	78·3
IF_7	g	−225·6	−195·6	82·8
K	c	0	0	15·2
K^+	g	123·0	—	—
K^+	aq	−60·04	−67·46	24·5
K_2O_2	c	−118	—	—
KO_2	c	−67	—	—
KN_3	c	0·3	—	—
KHF_2	c	−219·98	−203·73	24·92

Selected thermodynamic data at 25°

Compound	State	ΔH_f^0 (kcal/mole)	ΔG_f^0 (kcal/mole)	S^0 (cal/deg.mole)
Li	c	0	0	6·70
Li^+	g	164·3	—	—
Li^+	aq	$-66·55$	$-70·22$	3·4
Li_2O	c	$-142·4$	$-133·8$	8·97
Li_2O_2	c	$-151·7$	—	—
LiN_3	c	2·6	—	—
Li_2CO_3	c	$-290·54$	$-270·66$	21·60
Mg	c	0	0	7·77
Mg^{2+}	g	561·3	—	—
Mg^{2+}	aq	$-110·41$	$-108·99$	$-28·2$
MgO	c	$-143·77$	$-136·12$	6·6
MgO_2	c	$-148·9$	—	—
$Mg(OH)_2$	c	$-221·0$	$-199·27$	15·09
$Mg(NO_3)_2$	c	$-188·72$	$-140·63$	39·2
$MgCO_3$	c	-266	-246	15·7
Mn	α–c	0	0	7·59
Mn^{2+}	g	604	—	—
Mn^{2+}	aq	$-53·6$	$-55·0$	-19
MnO	c	$-92·05$	$-86·8$	14·3
MnO_4^-	aq	$-129·7$	$-107·1$	45·4
$MnCO_3$	c	$-213·9$	$-195·4$	20·5
MnF_2	c	-190	-180	22·2
$MnBr_2$	c	-91	-89	36·0
MnI_2	c	-59	-60	38·0
N_2	g	0	0	45·77
N	g	112·979	108·883	36·622
N_3^-	g	43·2	—	—
NO	g	21·57	20·69	50·347
NO^+	g	236·4	—	—
NO_2	g	7·93	12·26	57·35
NO_3^-	aq	$-49·56$	$-26·61$	35·0
N_2O	g	19·61	24·90	52·52
N_2O_3	g	20·01	33·32	74·61
N_2O_4	l	$-4·66$	23·29	50·0
N_2O_4	g	2·19	23·38	72·70
N_2O_5	c	$-10·3$	27·2	42·6
N_2O_5	g	2·7	27·5	85·0
NH_3	g	$-11·02$	$-3·94$	45·97
N_2H_4	l	12·10	35·67	28·97
N_2H_4	g	22·80	38·07	56·97
NH_4^+	aq	$-31·67$	$-18·97$	27·1
HNO_2	aq	$-28·5$	$-13·3$	36·5
NH_2OH	c	$-27·3$	—	—
NF_3	g	$-29·8$	$-20·0$	62·29

Compound	State	ΔH_f^0 (kcal/mole)	ΔG_f^0 (kcal/mole)	S^0 (cal/deg.mole)
N_2F_4	g	$-1\cdot7$	$19\cdot4$	$71\cdot96$
NCl_3	l	55	—	—
NOF	g	$-15\cdot9$	$-12\cdot2$	$59\cdot27$
NOCl	g	$12\cdot36$	$15\cdot78$	$62\cdot52$
NH_4F	c	$-110\cdot89$	$-83\cdot36$	$17\cdot20$
NH_4Cl	c	$-75\cdot15$	$-48\cdot51$	$22\cdot6$
NH_4Br	c	$-64\cdot73$	$-41\cdot9$	27
NH_4I	c	$-48\cdot14$	$-26\cdot9$	28
Na	c	0	0	$12\cdot2$
Na^+	g	$145\cdot9$	—	—
Na^+	aq	$-57\cdot279$	$-62\cdot589$	$14\cdot4$
Na_2O	c	$-99\cdot4$	$-90\cdot0$	$17\cdot4$
Na_2O_2	c	$-120\cdot6$	—	—
NaO_2	c	$-61\cdot9$	—	—
Na_2CO_3	c	$-270\cdot3$	$-250\cdot4$	$32\cdot5$
$NaHF_2$	c	$-216\cdot6$	—	—
Ni	c	0	0	$7\cdot20$
Ni^{2+}	g	700	—	—
Ni^{2+}	aq	$-12\cdot8$	$-11\cdot1$	-30
O_2	g	0	0	$48\cdot996$
O	g	$59\cdot553$	$55\cdot388$	$38\cdot467$
O^-	g	$24\cdot29$	—	—
O_2^+	g	$281\cdot48$	—	—
O_3	g	$34\cdot1$	$39\cdot0$	$57\cdot08$
P	white	0	0	$9\cdot82$
P	red	$-4\cdot2$	$-2\cdot9$	$5\cdot45$
P	g	$79\cdot8$	—	—
P_2	g	$34\cdot5$	$24\cdot8$	$52\cdot108$
P_4	g	$14\cdot08$	$5\cdot85$	$66\cdot89$
PO_4^{3-}	aq	$-305\cdot9$	$-244\cdot0$	-53
P_4O_6	c	$-392\cdot0$	—	—
P_4O_{10}	hexagonal	$-713\cdot2$	$-644\cdot8$	$54\cdot70$
PH_3	g	$1\cdot3$	$3\cdot2$	$50\cdot22$
P_2H_4	g	$5\cdot0$	—	—
H_3PO_3	c	$-230\cdot5$	—	—
H_3PO_3	aq	$-230\cdot6$	—	—
PF_3	g	$-219\cdot6$	$-214\cdot5$	$65\cdot28$
PCl_3	l	$-76\cdot4$	$-65\cdot1$	$51\cdot9$
PCl_3	g	$-68\cdot6$	$-64\cdot0$	$74\cdot49$
PBr_3	g	$-33\cdot3$	$-38\cdot9$	$83\cdot17$
PF_5	g	$-381\cdot4$	—	—
PCl_5	g	$-89\cdot6$	$-73\cdot0$	$87\cdot11$
PH_4Cl	c	$-34\cdot7$	—	—
PH_4Br	c	$-30\cdot5$	$-11\cdot4$	$26\cdot3$
PH_4I	c	$-16\cdot7$	$0\cdot2$	$29\cdot4$

Selected thermodynamic data at 25°

Compound	State	ΔH_f^0 (kcal/mole)	ΔG_f^0 (kcal/mole)	S^0 (cal/deg.mole)
Pb	c	0	0	15·51
Pb²⁺	g	567·4	—	—
PbO	red	−52·40	−45·25	16·2
PbF₂	c	−158·5	−148·1	29
PbCl₂	c	−85·7	−74·9	32·6
PbI₂	c	−41·6	−41·3	42·3
PbCO₃	c	−167·3	−149·7	31·3
Ra	c	0	0	17
Ra²⁺	aq	−126	−134·5	13
RaO	c	−125	−117·5	—
Ra(NO₃)₂	c	−237	−190·3	52
RaSO₄	c	−352	−326·0	34
Rb	c	0	0	16·6
Rb⁺	g	117·4	—	—
Rb⁺	aq	−59·4	−67·45	28·0
Rb₂O₂	c	−101·7	—	—
RbO₂	c	−63·1	—	—
RbN₃	c	−0·1	—	—
RbHF₂	c	−217·8	—	—
S	rhombic	0	0	7·60
S	g	66·6	56·9	40·094
S²⁻	aq	7·9	20·5	−3·5
S₂	g	30·68	18·90	54·51
S₄²⁻	aq	5·5	16·5	24·7
S₅²⁻	aq	5·1	15·7	33·6
S₈	g	24·45	11·87	102·98
SO	g	1·5	−4·7	53·02
SO₂	g	−70·944	−71·749	59·30
SO₃	β–c	−108·63	−88·19	12·5
SO₃	g	−94·58	−88·69	61·34
SO₄²⁻, H₂SO₄	aq	−217·32	−177·97	4·8
H₂S	g	−4·93	−8·02	49·16
H₂S₄	l	−7·85	—	—
H₂S₆	l	−8·85	—	—
SCl₂	g	−4·7	—	—
SF₄	g	−185·2	−174·8	69·77
SF₆	g	−291·8	−266·9	69·72
Sb	c	0	0	10·92
Sb	g	62·7	53·1	43·06
Sb₄	g	49·0	33·8	84
SbH₃	g	34·681	35·31	55·61
SbCl₃	c	−91·34	−77·37	44·0
SbCl₃	g	−75·0	−72·0	80·71
SbBr₃	g	−46·5	−53·5	89·09

Compound	State	ΔH_f^0 (kcal/mole)	ΔG_f^0 (kcal/mole)	S^0 (cal/deg.mole)
Se	black	0	0	10·144
Se	g	49·2	39·6	42·22
H_2Se	g	7·1	3·8	52·32
$SeCl_2$	g	−7·6	—	—
Se_2Cl_2	g	4	—	—
Se_2Br_2	g	7	—	—
Si	c	0	0	4·50
Si	g	108·9	98·3	40·12
Si_2	g	142	128	54·92
SiO	g	−23·8	−30·2	50·55
SiO_2	quartz	−217·72	−204·75	10·00
SiO_2	g	−77	—	—
SiH_4	g	8·2	13·6	48·88
Si_2H_6	g	19·2	30·4	65·14
Si_3H_8	g	28·9	—	—
SiF_4	g	−386·0	−375·9	67·49
$SiCl_4$	l	−164·2	−148·2	57·3
$SiCl_4$	g	−157·0	−147·5	79·02
$SiBr_4$	g	−99·3	−103·2	90·29
Si_3N_4	c	−177·7	−153·6	24·2
SiC	cubic	−15·6	−15·0	3·97
Sn	white	0	0	12·32
Sn	gray	−0·50	0·03	10·55
Sn	g	72·2	63·9	40·24
SnH_4	g	38·9	45·0	54·39
$SnCl_4$	g	−112·7	−103·3	87·4
$SnBr_4$	g	−75·2	−79·2	98·43
$Sn(CH_3)_4$	g	−4·5	—	—
Sr	c	0	0	13·0
Sr^{2+}	g	427·7	—	—
Sr^{2+}	aq	−130·38	−133·2	−9·4
SrO	c	−141·1	−133·8	13·0
SrO_2	c	−153·6	—	—
$Sr(OH)_2$	c	−229·3	−207·8	—
$Sr(NO_3)_2$	c	−233·3	−186	—
$SrCO_3$	c	−291·9	−271·9	23·2
Te	c	0	0	11·88
Te	g	45·5	36·1	43·65
H_2Te	g	23·8	—	—
TeF_6	g	−327·2	—	—
Ti	c	0	0	7·24
$TiCl_2$	c	−123	−112	24·3
$TiCl_3$	c	−172·2	−156·2	33·4

Selected thermodynamic data at 25°

Compound	State	ΔH_f^0 (kcal/mole)	ΔG_f^0 kcal/mole)	S^0 (cal/deg.mole)
Tl	c	0	0	15·34
Tl^{2+}	g	658·4	—	—
TlCl	c	−48·79	−44·20	26·59
V	c	0	0	7·05
VCl$_2$	c	−107	−96	23·2
VCl$_3$	c	−138	−121	31·3
Xe	g	0	0	40·5290
Xe$^+$	g	281·2	—	—
XeF$_2$	g	−25·9	—	—
XeF$_4$	g	−51·5	—	—
XeF$_6$	g	−70·4	—	—
XeO$_3$	c	96	—	—
Zn	c	0	0	9·95
Zn^{2+}	g	665·1	—	—
Zn^{2+}	aq	−36·6	−35·2	−26·0
ZnO	c	−83·25	−76·13	10·5
ZnF$_2$	c	−182·7	−170·5	17·61
ZnCO$_3$	c	−194·2	−174·8	19·7

A few additional values are recorded in tables 1.1, 2.11, 3.1, 3.2, 3.3, 3.4, and 4.1.

Index